Susanne Adams

Monitoring of thin sea ice within polynyas using MODIS data

Susanne Adams

Monitoring of thin sea ice within polynyas using MODIS data

Südwestdeutscher Verlag für Hochschulschriften

Impressum / Imprint
Bibliografische Information der Deutschen Nationalbibliothek: Die Deutsche Nationalbibliothek verzeichnet diese Publikation in der Deutschen Nationalbibliografie; detaillierte bibliografische Daten sind im Internet über http://dnb.d-nb.de abrufbar.
Alle in diesem Buch genannten Marken und Produktnamen unterliegen warenzeichen-, marken- oder patentrechtlichem Schutz bzw. sind Warenzeichen oder eingetragene Warenzeichen der jeweiligen Inhaber. Die Wiedergabe von Marken, Produktnamen, Gebrauchsnamen, Handelsnamen, Warenbezeichnungen u.s.w. in diesem Werk berechtigt auch ohne besondere Kennzeichnung nicht zu der Annahme, dass solche Namen im Sinne der Warenzeichen- und Markenschutzgesetzgebung als frei zu betrachten wären und daher von jedermann benutzt werden dürften.

Bibliographic information published by the Deutsche Nationalbibliothek: The Deutsche Nationalbibliothek lists this publication in the Deutsche Nationalbibliografie; detailed bibliographic data are available in the Internet at http://dnb.d-nb.de.
Any brand names and product names mentioned in this book are subject to trademark, brand or patent protection and are trademarks or registered trademarks of their respective holders. The use of brand names, product names, common names, trade names, product descriptions etc. even without a particular marking in this works is in no way to be construed to mean that such names may be regarded as unrestricted in respect of trademark and brand protection legislation and could thus be used by anyone.

Coverbild / Cover image: www.ingimage.com

Verlag / Publisher:
Südwestdeutscher Verlag für Hochschulschriften
ist ein Imprint der / is a trademark of
OmniScriptum GmbH & Co. KG
Heinrich-Böcking-Str. 6-8, 66121 Saarbrücken, Deutschland / Germany
Email: info@svh-verlag.de

Herstellung: siehe letzte Seite /
Printed at: see last page
ISBN: 978-3-8381-3856-5

Zugl. / Approved by: Trier, Universtiät Trier, Dissertation, 2012

Copyright © 2014 OmniScriptum GmbH & Co. KG
Alle Rechte vorbehalten. / All rights reserved. Saarbrücken 2014

Contents

Abstract IV
Zusammenfassung V
1   **Introduction** 1
    1.1   Motivation 1
    1.2   Polynyas 3
    1.3   Objectives and organization of the thesis 6
2   **Remote sensing of thin sea ice** 8
    2.1   Sensor systems 8
        2.1.1   Optical sensor systems 8
        2.1.2   Passive microwave sensors 9
    2.2   Review of methods to derive sea-ice and polynya characteristics 10
        2.2.1   Ice-surface temperature 11
        2.2.2   Thermal-infrared thin-ice thickness 12
        2.2.3   Thermal-infrared sea-ice concentration 13
        2.2.4   Passive microwave sea-ice concentration 14
        2.2.5   Passive microwave thin-ice thickness 16
        2.2.6   Polynya signature simulation method 20
        2.2.7   Methods used in this thesis 21
3   **Data sets** 22
    3.1   Satellite remote sensing data sets 23
        3.1.1   Optical remote sensing data 23
        3.1.2   Passive microwave data 24
        3.1.3   Envisat ASAR 24
    3.2   In-situ data sets 25
    3.3   Model data sets 29
        3.3.1   Atmospheric model data sets 29
            3.3.1.1   NCEP 29
            3.3.1.2   COSMO 30
        3.3.2   Sea-ice/ocean model data set 31
4   **Retrieval of MODIS thin-ice thickness in the Laptev Sea** 33
    4.1   Verification of MODIS ice-surface temperature 33

I

4.2	Thin-ice thickness algorithm	36
4.3	Improvement of the atmospheric flux calculations	42
4.4	Sensitivity analysis	45
4.4.1	Statistical sensitivity analysis	45
4.4.1.1	Method	45
4.4.1.2	Results and discussion	46
4.4.2	Comparison of ice-thickness data sets using different atmospheric data	49
4.4.3	Nearly-coincident MODIS T_s and TIT as an uncertainty indicator	56
4.4.4	Summary and conclusion of the sensitivity analyses	59
4.5	Transfer of the thin-ice thickness algorithm to other regions	61
4.5.1	Lincoln Sea (Arctic)	61
4.5.2	Weddell Sea (Antarctic)	67
4.6	Retrieval of further MODIS products	71
4.6.1	Daily thin-ice thickness maps	71
4.6.2	Monthly fast-ice masks	77
4.7	Intercomparison of various remote sensing data sets	81
4.7.1	Results	81
4.7.2	Conclusions	85
5	**Verification of numerical models using remote sensing data**	**87**
5.1	Sea-ice/ocean models and the efforts to simulate or prescribe fast ice	87
5.2	Evaluation of simulated sea-ice concentration	89
5.2.1	Sea-ice concentration data sets	89
5.2.2	Evaluation area and variables	90
5.2.3	Open-water area	90
5.2.4	Polynya area	93
5.2.5	A case study	96
5.2.6	Summary and discussion	98
6	**Concluding remarks**	**102**
6.1	Conclusions	102
6.2	Outlook	103
Bibliography		**106**
List of Symbols		**118**
List of Abbreviations		**120**
Appendix		**122**

Acknowledgements **129**

Abstract

Arctic and Antarctic polynya systems are of high research interest since extensive new ice formation takes place in these regions. The monitoring of polynyas and the ice production is crucial with respect to the changing sea-ice regime. The thin-ice thickness (TIT) distribution within polynyas controls the amount of heat that is released to the atmosphere and has therefore an impact on the ice-production rates.

This thesis presents an improved method to retrieve thermal-infrared thin-ice thickness distributions within polynyas. TIT with a spatial resolution of 1 km × 1 km is calculated using the MODIS ice-surface temperature and atmospheric model variables within the Laptev Sea polynya for the winter periods 2007/08 and 2008/09.

The improvement of the algorithm is focused on the surface-energy flux parameterizations. Furthermore, a thorough sensitivity analysis is applied to quantify the uncertainty in the thin-ice thickness results. An absolute mean uncertainty of ±4.7 cm for ice below 20 cm of thickness is calculated. Furthermore, advantages and drawbacks using different atmospheric data sets are investigated.

Daily MODIS TIT composites are computed to fill the data gaps arising from clouds and shortwave radiation. The resulting maps cover on average 70 % of the Laptev Sea polynya. An intercomparison of MODIS and AMSR-E polynya data indicates that the spatial resolution issue is essential for accurately deriving polynya characteristics.

Monthly fast-ice masks are generated using the daily TIT composites. These fast-ice masks are implemented into the coupled sea-ice/ocean model FESOM. An evaluation of FESOM sea-ice concentrations is performed with the result that a prescribed high-resolution fast-ice mask is necessary regarding the accurate polynya location. However, for a more realistic simulation of other small-scale sea-ice features further model improvements are required.

The retrieval of daily high-resolution MODIS TIT composites is an important step towards a more precise monitoring of thin sea ice and sea-ice production. Future work will address a combined remote sensing – model assimilation method to simulate fully-covered thin-ice thickness maps that enable the retrieval of accurate ice production values.

Zusammenfassung

Arktische und antarktische Polynja-Systeme agieren als Gebiete intensiver Eisproduktion und stehen daher im Fokus der Meereisforschung. Aufgrund der sich ändernden Meereisbedingungen in den Polarregionen ist die Beobachtung von Polynjen von großem Interesse. Die Verteilung des dünnen Eises in Polynjen kontrolliert den Wärmeverlust zur Atmosphäre und damit die Eisproduktionsraten.

Diese Arbeit beschäftigt sich mit der Ableitung von thermalen Dünneisdicken in Polynjen. Dünneisdicken mit einer räumlichen Auflösung von 1 km × 1 km werden auf Basis von MODIS Eisoberflächentemperaturen und atmosphärischen Modelldaten für die Laptev See Polynja berechnet. Der Datensatz wird für die beiden Winter 2007/08 und 2008/09 bereitgestellt.

Ein bestehender Algorithmus wird bezüglich der Parametrisierung von Energieflüssen an der Oberfläche verbessert. Weiterhin wird eine ausführliche Sensitivitätsanalyse durchgeführt, um den Fehler in den Dünneisdicken zu quantifizieren. Der mittlere absolute Fehler beträgt ±4.7 cm für Eisdicken kleiner als 20 cm. Die Vor- und Nachteile der Verwendung von unterschiedlichen atmosphärischen Datensätzen werden untersucht.

Nach der Sensitivitätsanalyse werden tägliche MODIS Dünneisdicken-Komposite erstellt, um die Lücken, die durch Wolken und kurzwellige Einstrahlung entstehen, zu füllen. Die Komposite decken im Mittel 70 % der Laptev See Polynja ab. Ein Vergleich von MODIS und AMSR-E Datensätzen zeigt, dass die räumliche Auflösung wesentlich ist, um Polynja-Charakteristika hinreichend genau abzuleiten.

Monatliche Festeismasken werden aus den täglichen Dünneisdicken-Kompositen abgeleitet und in das Meereis-Ozeanmodell FESOM implementiert. Die Verifizierung von simulierten Meereiskonzentrationen belegt, dass eine Festeis-Parametrisierung erforderlich ist, um die Lage der Polynja realistisch wiederzugeben. Es sind jedoch weitere Modellverbesserungen notwendig, um die realistische Simulation weiterer kleinskaliger Meereiseigenschaften zu gewährleisten.

Die Ableitung täglicher, hochaufgelöster MODIS Dünneisdicken ist ein wichtiger Schritt in Richtung einer genaueren Beobachtung von dünnem Meereis und der Meereisproduktion. Zukünftige Arbeiten werden sich mit der Assimilierung von MODIS Dünneidicken in das Meereis-Ozeanmodell FESOM beschäftigen, um Dünneisdicken-Karten mit kompletter Abdeckung zu simulieren und damit die Berechnung von realistischen Eisproduktionen zu ermöglichen.

1 Introduction

1.1 Motivation

The Arctic climate system reacts quickly and sensitive to global warming. Recent studies show that the warming due to greenhouse gases in the Arctic is twice as high as in the mid-latitudes (Serreze and Francis [2006], Screen and Simmonds [2010]). This phenomenon is generally referred to as Arctic amplification and might be partially caused by the decreasing sea-ice extent and thickness as well as changes in cloud cover and atmospheric water vapor content (Screen and Simmonds [2010], Serreze et al. [2009], Lu and Cai [2009], Graversen and Wang [2009]).

The impact of climate change on sea-ice extent and thickness has been investigated in several studies. According to Comiso et al. [2008], the sea-ice decline has accelerated in the last decade. The trend of decreasing sea-ice extent shifted from -2.2 % in 1979-1996 to -10.1 % in 1997-2007. The summer sea-ice minimum was reached in September 2007 with a sea-ice extent of 4.1×10^6 km^2, 37 % lower than the climatological average (Comiso et al. [2008]). In association with the shrinking sea-ice extent, a thinning of sea ice is recorded (Giles et al. [2008], Kwok and Rothrock [2009]). For the winter season 2007/08, Giles et al. [2008] estimated an ice-thickness anomaly of 26 cm below the six-year mean from 2002/03-2007/08. In addition to the observation of the large-scale sea-ice extent and thickness, small scale-features must also be investigated with respect to climate change. In particular, significant attention must be paid to the polynya systems within the marginal sea-ice zone of the Arctic. Polynyas are non-linear areas of open water and thin sea ice enclosed by thick ice (Morales Maqueda et al. [2004]). These features contribute substantially to the Arctic sea-ice budget through extensive wintertime ice production rates (Morales Maqueda et al. [2004]). The evolution process and the characteristics of polynyas are described in Section 1.2.

The Laptev Sea is internationally designated as a key region to investigate the climate change within the Arctic shelf seas (Arctic Climate Assessment (ACIA) [2004]). This shelf sea is located between the Severnaya Zemlya in the west, the Lena Delta in the south and the New Siberian Islands in the east (Figure 1).

A particularity of the Laptev Sea is the huge freshwater inflow of 750 km^3 per year (Rigor and Colony [1997]). 70 % of the freshwater inflow discharging into the Laptev Sea originates from the Lena River. Yenisei, Khatanga and Anabar are other important rivers discharging into the Laptev Sea. The freshwater inflow affects the stratification of the water layers as well as the salinity level.

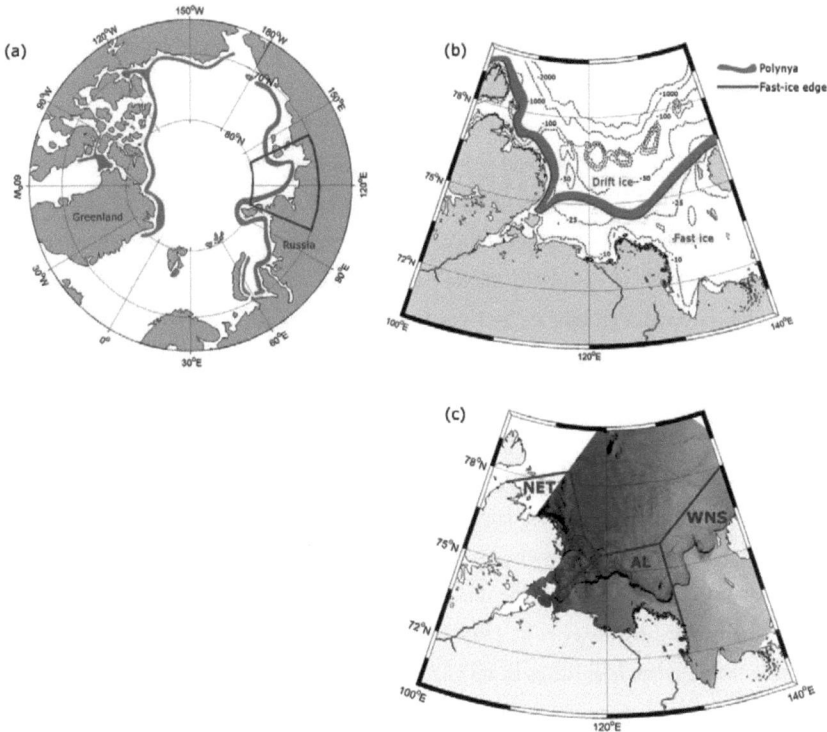

Figure 1: (a) Overview map of the Arctic. The blue band along the coastline denotes the polynya system occurring during the freezing period in the Arctic. The red box marks the Laptev Sea in the Siberian Arctic shown in (b). (b) Map of the Laptev Sea. The dashed black lines are the bathymetric contour lines. The red line denotes schematically the average fast-ice edge. The blue band shows the regions where polynyas occur in the Laptev Sea. (c) MODIS channel 1 image of the Laptev Sea for 22 April, 2008 1055 UTC. Polynyas can be seen as the dark narrow band along the fast-ice edge. Red boxes denote the Laptev Sea polynya subsets: the northeastern Taimyr polynya (NET), the Taimyr polynya (T), the Anabar-Lena polynya (AL) and the western New Siberian polynya (WNS). All subsets together are called LAP.

During the freezing period (October-June), fast ice forms along the coastline of the Laptev Sea and reaches its maximum extent in April. Off-shore wind conditions enable the opening of polynyas along the fast-ice edge during the winter season. Huge amounts of new ice and dense water are produced within the Laptev Sea polynya (Dmitrenko et al. [2005]). Due to the high new ice formation rates the Laptev Sea is called 'ice factory'. This newly formed ice is a major source of sea ice that is transported through the Transpolar Drift (Dethleff et al. [1998]). Between 1979 and 1995, the average ice outflow of the Laptev Sea was 483,000 km^2 (Alexandrov et al. [2000]). The annual outflow varied between 251,000 km^2 in the winter period 1984/85 and 732,000 km^2 in 1988/89.

Previous studies addressing the sea-ice production in the Laptev Sea show large discrepancies in the amount of newly formed ice that might be, at least partly, attributed to methodological differences (Dethleff et al. [1998], Winsor and Björk [2000], Dmitrenko et al. [2009], Willmes et al. [2011], Tamura and Oshima [2011]). For instance, Dethleff et al. [1998] used the simple relationship between wind direction/speed and polynya area to derive an ice production of 258 km^3 in the Laptev Sea polynya for the winter season 1991/1992. In contrast, Willmes et al. [2011] calculated for the same winter an ice production of 63 km^3 if the polynya is covered with thin ice and 114 km^3 if the polynya is ice free. They used passive microwave and thermal-infrared satellite data in combination with atmospheric reanalysis data to retrieve the ice production. Due to the large differences in the ice-production values, it is necessary to improve the existing methods to get more accurate results. According to Willmes et al. [2011], the knowledge of the thin-ice thickness distribution within the polynyas needs to be refined because the impact of a thin-ice layer on the heat loss is crucial. The previous methods mostly assume a completely ice-free polynya or an empirical thin-ice distribution. The importance of the thin-ice thickness distribution for the ice production was also demonstrated by Ebner et al. [2011]. They suggest that a thin-ice layer of around 5 cm reduces the turbulent heat fluxes by up to 270 W m^{-2} depending on the temperature conditions and the wind speed of the atmospheric boundary layer above the thin ice.

According to these considerations, the overall aim of this thesis is the implementation of an improved thin-ice thickness algorithm to monitor the ice-thickness distribution and variability of polynyas in the Laptev Sea. A data set of nearly fully-covered high-resolution thin-ice thickness maps based on remote sensing data will be provided.

1.2 Polynyas

Polynyas are recurring elongated areas of open water and thin ice occurring along the coastline or the fast-ice edge within the marginal sea-ice zones of the polar oceans (Smith et al. [1990], Martin [2001]). These features can be '[...] tens to tens of thousands of square kilometers in areal extent [...]' and appear '[...] at locations where a more consolidated and thicker ice cover would be climatologically expected' (Barber and Massom [2007]).

There are two different mechanisms controlling the formation of polynyas: (1) wind-driven polynya formation and (2) polynya formation due to convective ocean currents (Figure 2a). The latent heat polynya opens during off-shore wind conditions when the ice is advected away from the coast or the fast-ice edge (Barber and Massom [2007]). Because the sea water is at its freezing point, instantaneous formation of frazil ice occurs when the insulating consolidated ice cover is removed. During this process latent and sensible heat is released into the atmosphere, brine is rejected and dense water is produced. Continuing off-shore wind pushes the new ice seawards, causing continued ice production. Several synonyms for latent heat polynya are found in the literature: shelf-water polynya, coastal polynya and wind-driven flaw polynya.

Figure 2b presents a photograph of the Laptev Sea polynya. This photograph shows a narrow open-water area and different new-ice types. According to the World Meteorological Organization [1990], the new-ice types are called frazil ice, grease ice and nilas. Frazil ice consists of newly formed, loose ice crystals. In the photograph, the frazil ice is collected along the downwelling zones of the Langmuir circulation (Figure 2b). Ice crystals that clump together to form a soupy layer are called grease ice. Nilas are areas of consolidated ice up to 10 cm thick. The World Meteorological Organization [1990] describes nilas as a 'thin elastic crust of ice' and distinguishes between dark and light nilas. Dark nilas is thinner than 5 cm of ice thickness and is very dark in color. Areas of ice with a thickness between 5 and 10 cm are classified as light nilas and have a lighter color. The photograph shows an area of dark nilas in proximity of the fast-ice edge (Figure 2b). Ice with a thickness between 10 and 30 cm is called young ice (World Meteorological Organization [1990]).

On the on-shore side, the polynya is bordered by the coastline or the fast-ice edge. Fast ice is sea ice that is contiguous to a shore and does not move with ocean currents or winds (World Meteorological Organization [1990], Mahoney et al. [2007]). The photograph shows snow-covered fast ice in the foreground (Figure 2b). On the off-shore side, drift ice is attached to the polynya. The definition of drift ice is not specified by age, form, origin or thickness but it has the characteristic to move or drift with winds, ocean currents and tides (World Meteorological Organization [1990]). In the background, the photograph shows a small strip of snow-covered drift ice (Figure 2b).

The second polynya type is the sensible heat polynya (Figure 2a). This polynya forms due to convective ocean currents Barber and Massom [2007]. Warm upwelling water melts the ice and an open-water area occurs. The closing process is initiated by winds that advect ice into the polynya. The size and persistence of the sensible heat polynya result from the interaction between the upwelling warm water and the advection of ice into the polynya.

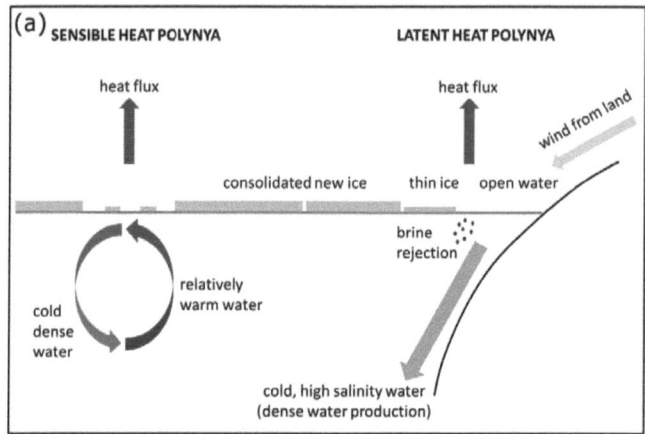

Figure 2: (a) Scheme showing the formation of the two polynya types: sensible heat polynya and latent heat polynya. (b) Photograph of the Laptev Sea polynya (latent heat polynya) with different ice types. © Photograph: T. Ernsdorf, University of Trier (2008).

Latent heat polynyas are important for the Arctic sea-ice system because these features are areas of substantial new ice formation. With respect to climate change it is essential to monitor the ice production within polynyas. The key variables required to derive the ice production within polynyas are the polynya area, the thin-ice thickness distribution within the polynya and atmospheric quantities (e.g., 2-m temperature, wind speed).

Due to the good spatial and temporal availability, satellite observations provide valuable data sets from which to derive polynya area and thin-ice thickness. Both optical and passive microwave sensors provide data that is appropriate for the observation of polynyas. The Moderate Resolution Imaging Spectroradiometer (MODIS) and the Advanced Very High Resolution Radiometer (AVHRR) detect visible and infrared radiation. These sensors have a considerably finer spatial resolution than the passive microwave sensors, but the data is only useful under clear-sky conditions. The two major passive microwave sensors are the Special Sensor Microwave Imager (SSM/I) and the Advanced Microwave Scanning Radiometer – Earth Observation System (AMSR-E). Using the remote sensing data sets a variety of methods are developed to derive polynya area and thin-ice thickness distribution.

Atmospheric variables are provided by atmospheric models or local measurements.

1.3 Objectives and organization of the thesis

The aim of this thesis is the improvement of the thin-ice thickness monitoring within polynyas to enable a more accurate calculation of the polynya ice production. The state-of-the-art of polynya investigation shows that recently implemented methods allow the observation of polynya dynamics and the polynya ice production. However, discrepancies in the results indicate that further research is required in order to gain improved data sets to aid in the monitoring of polynyas. Hence, the objectives of this thesis are:

(1) the improvement of the thermal-infrared thin-ice thickness retrieval;

(2) an uncertainty estimation of the satellite-based thin-ice thickness data sets;

(3) the computation of daily MODIS thin-ice thickness composites;

(4) the allocation of MODIS fast-ice masks for prescription in sea-ice/ocean models.

In order to fulfill the objectives, at first an established thin-ice thickness retrieval is modified by improving its parameterizations. The parameterizations of surface-energy fluxes in previous thin-ice thickness retrieval methods are not state-of-the-art for the atmospheric boundary layer, and therefore more improvements are required. Secondly, a sensitivity analysis of the thin-ice thickness algorithm is performed. The uncertainty estimation includes a statistical and comparative part to quantify the uncertainty of the thin-ice thickness data set and to determine the impact of the input variables on the uncertainty in the thin-ice thickness. After the assessment of the method and the input data sets, a daily MODIS thin-ice thickness composite of the Laptev Sea is produced for two winter seasons and compared to other remote sensing data sets. Based on the daily MODIS thin-ice thickness maps, monthly fast-ice masks are derived. The polynya simulation of a coupled sea-ice/ocean model is improved by the prescription of the MODIS fast-ice area. The sea-ice concentrations simulated by this model are evaluated using remote sensing data.

The thesis is organized as follows: After the introduction, a review of the methods used for the remote sensing of thin sea ice is given. The sensor systems and methods to derive sea ice quantities are described. In the subsequent section the used data sets are introduced. Section 4 deals with the thin-ice thickness algorithm and its improvement. Furthermore, a thorough sensitivity analysis is provided, the algorithm is applied to another shelf sea in the Arctic and to one in the Antarctic, and subsequent processing of MODIS thin-ice thickness maps is described. After the comprehensive analysis of the thin-ice thickness algorithm, Section 5 addresses the treatment of fast ice in numerical models and the evaluation of simulated sea-ice concentrations with respect to a fast-ice prescription. The last section concludes the findings of the thesis and provides an outlook on future work.

2 Remote sensing of thin sea ice

2.1 Sensor systems

2.1.1 Optical sensor systems

Various optical and passive microwave sensor systems monitor the thin-ice regions in the Laptev Sea polynya. The basic specifications and improvements of these sensors since the 1970s are reviewed in this section.

Optical satellite data has been applied for polynya investigations for more than 30 years. In October, 1978 the TIROS-N satellite was launched with the first Advanced Very High Resolution Radiometer (AVHRR) on board. This imaging radiometer captures four wide channels from visible to thermal-infrared wavelengths. The second AVHRR sensor, equipped with five spectral channels, began operating on board the NOAA-7 satellite in 1981. The third generation of the AVHRR sensor is equipped with six spectral channels (0.58 – 12.5 µm) and has collected data since 1998, initially on board the NOAA-15 satellite. All AVHRR channels have a spatial resolution of 1.1 km × 1.1 km at nadir. At least two AVHRR instruments are operating at the same time to guarantee global coverage twice-daily.

The successor of AVHRR is the Moderate Resolution Imaging Spectroradiometer (MODIS) with finer spatial and spectral resolution. MODIS is an imaging radiometer that measures electromagnetic radiation from 0.4 µm (visible) to 14.4 µm (thermal infrared). 36 discrete spectral bands with a spatial resolution from 0.25 km × 0.25 km to 1 km × 1 km capture this wavelength range (Table 1). The MODIS sensor has been deployed on two satellites, Terra and Aqua, which were launched in December, 1999 and May, 2002, respectively. The satellites are positioned in a sun-synchronous polar orbit at an altitude of 705 km and an orbital period of roughly 100 minutes. Making 14.4 orbits per day, the two MODIS instruments are able to scan the whole earth within two days (Barnes et al. [1998], Hall et al. [2004]).

Table 1: Basic specifications of the MODIS sensor.

Feature	MODIS
Launch	first instrument 1999 (Terra) second instrument 2002 (Aqua)
Orbit	sun-synchronous, polar
Altitude	705 km
Zenith angle	55°
Swath size	2330 km
Spectral range	0.4 µm (visible) to 14.4 µm (thermal infrared)
Spatial resolution	0.25 km × 0.25 km (bands 1-2)
	0.50 km × 0.50 km (bands 3-7)
	1.00 km × 1.00 km (bands 8-36)
Data sets obtained by others	MOD/MYD29 ice-surface temperature (T_s) obtained by NASA
	MODIS sea-ice concentration (SIC) obtained by S. Willmes, University of Trier
Own data sets	MODIS thin-ice thickness (TIT)

2.1.2 Passive microwave sensors

Passive microwave data has been available since 1973. The first data set was provided by the Electrically Scanning Microwave Radiometer (ESMR) on the Nimbus-5 satellite from 1973 to 1976. This instrument measured at a single polarization and a frequency of 19 GHz with a field of view (FOV) of 25 km × 25 km at nadir and 160 km × 40 km at scan extremes (Parkinson et al. [1987]). The Scanning Multichannel Microwave Radiometer (SMMR) was deployed on Nimbus-7 in 1978 and was equipped with five channels between 7 GHz and 37 GHz measuring the brightness temperature at horizontal and vertical polarizations (Cavalieri et al. [1984], Gloersen et al. [1984]). The size of the FOV varies between 171 km × 157 km (7 GHz) and 35 km × 34 km (37 GHz). Since 1987, the Special Sensor Microwave Imager (SSM/I) has been on board the Defense Meteorological Satellite Program (DMSP) satellites. SSM/I detects the brightness temperature at four frequencies between 19 GHz and 85 GHz. Three of the frequencies (19, 37 and 85 GHz) are sampled in horizontal and vertical polarization; the 22 GHz frequency is only sampled vertically polarized (Hollinger et al. [1990], Hollinger et al. [1987]). In May, 2002 the Aqua satellite was launched carrying the Advanced Microwave Scanning Radiometer – Earth Observing System (AMSR-E). Table 2 gives an overview of the basic specifications of this sensor. The conical scanning microwave radiometer has 12 channels and operates at six frequencies (6.9, 10.7,

18.7, 23.8, 36.5 and 89 GHz) with horizontal and vertical polarization. The FOV is largest at 6.9 GHz (43 km × 75 km) and smallest at 89 GHz (4 km × 6 km). The gridded spatial resolution ranges from 25 km × 25 km to 6.25 km × 6.25 km. Due to a sensor failure, AMSR-E data is only available until September, 2011. It is replaced by AMSR-2 carried on the Shizuku satellite (GCOM-W1), which was launched in May, 2012. Operational data will be available in late 2012.

Despite the large time span from which SSM/I data is available (this data has been collected since 1987), only AMSR-E data is used in this thesis due to the finer spatial resolution of the AMSR-E sensor.

Table 2: Basic specifications of the AMSR-E sensor.

Feature	AMSR-E					
Launch	first instrument 2002 (Aqua), sensor failure in September 2011 second instrument 2012 (GCOM-W1)					
Orbit	sun-synchronous, polar					
Altitude	705 km					
Zenith angle	55°					
Swath size	1445 km					
Data sets obtained by others	AMSR-E / Aqua Daily Gridded Brightness temperature (T_B) obtained by NSIDC ASI sea-ice concentration (SIC) obtained by the University of Hamburg					
Own data sets	AMSR-E thin-ice thickness (TIT) AMSR-E polynya area retrieved by the polynya signature simulation method (PSSM)					
Frequencies (GHz)	6.9	10.7	18.7	23.8	36.5	89.0
Footprint size (km × km)	43 × 75	29 × 51	16 × 27	18 × 32	8 × 14	4 × 6
Gridded spatial resolution (km × km)	25 × 25	25 × 25	12.5 × 12.5	12.5 × 12.5	12.5 × 12.5	6.25 × 6.25

2.2 Review of methods to derive sea-ice and polynya characteristics

In this section, retrieval methods that characterize polynya quantities in Arctic polynyas are described. In the following, the different polynya quantities are explained:

- **Thin-ice thickness** (TIT): Different thin-ice types appear within a polynya (see Figure 2c; World Meteorological Organization [1990]). The definitions as to which ice thickness is called thin ice and hence classified as a part of the polynya differ in the literature. Here, ice up to a thickness of 20 cm is defined as thin ice. Previous studies (e.g., Willmes et al. [2010]) and the results of these studies show that thermal-infrared and passive microwave remote sensing data provide reliable information about the ice thickness in the range from 0 to 20 cm. Moreover, the distribution of ice thickness is required to calculate the ice production within polynyas. With respect to ice production, ice thickness up to 20 cm is sufficient since the heat loss and hence, the ice production is low in regions of thicker ice (Ebner et al. [2011]).

- **Sea-ice concentration** (SIC): All pixels in a satellite image that cover parts of the polar ocean are assumed to represent a mixture of open water and sea ice. Hence, the sea-ice concentration is the fraction of a pixel that is covered by sea ice. SIC ranges from 0 % (no ice is present within the pixel) to 100 % (the pixel is fully ice-covered).
- **Potential open water:** The potential open water is defined as 1 minus SIC.
- **Polynya area** (POLA): The number of pixels that are classified as a polynya multiplied by the pixel size. POLA can be defined: (1) by TIT less than 20 cm, (2) by all SIC pixels lower than a defined threshold (a threshold of 70 % SIC is used in this thesis), (3) by all T_s pixel higher than a temperature threshold and (4) by using the polynya signature simulation method (see Section 2.2.6).

2.2.1 Ice-surface temperature

Ice-surface temperature (T_s) is needed to derive the thin-ice thickness distribution within the polynya. The temperature is retrieved from thermal-infrared data with a split-window method (Hall et al. [2004]). The method is based on brightness temperatures T_B measured at 11 and 12 µm. These wavelengths are both located within atmospheric water vapor windows. At a wavelength of 11 µm approximately 80 % of the radiation upwelling from the surface is transmitted through the atmosphere. At a wavelength of 12 µm approximately 60 % of the radiation is transmitted because of the higher sensitivity to water vapor. Due to the different attenuation of the radiation in the two spectral bands the T_B difference contains information about the amount of water vapor. Hence, the application of the split window method is equivalent to an atmospheric water vapor correction. The simple regression model is defined as follows:

$$T_s = a + b \times T_{B,11} + c \times (T_{B,11} - T_{B,12}) + d \times [(T_{B,11} - T_{B,12}) \times (\sec\theta - 1)] \tag{1}$$

where $T_{B,11}$ and $T_{B,12}$ are the brightness temperatures at 11 and 12 µm, θ is the sensor scan angle providing information about the path length, and a-d are regression coefficients (Hall et al. [2004], Key et al. [1997]).

The regression coefficients a-d are determined using the low resolution radiative transfer model LOWTRAN developed for the prediction of atmospheric transmittance and background radiance Kneizys et al. [1988]. For this model, T_B at the sensor is simulated with radiosonde profiles. Then, a regression analysis is applied between the models' ice-surface temperature and the simulated T_B to determine the coefficients in Equation 1. The regression coefficients are calculated separately for the Arctic and Antarctic and for three temperature ranges (Hall et al. [2004], Riggs et al. [2012]).

In this thesis, the product of ice-surface temperatures derived by MODIS thermal-infrared data detected at the spectral bands 31 (11μm) and 32 (12 μm) is used Hall et al. [2007] (see Section 3.1.1).

2.2.2 Thermal-infrared thin-ice thickness

The retrieval of thermal-infrared thin-ice thickness (TIT) is based on the strong relation between ice-surface temperature and thin-ice thickness. TIT is calculated using the ice-surface temperature and additional atmospheric data with the assumption that the total atmospheric energy flux is balanced by the conductive heat flux through the ice.

First calculations of TIT were performed more than 30 years ago using airborne infrared imagery in combination with a very simple surface energy model (Kuhn et al. [1975]). Since the 1980s satellite thermal-infrared data has been provided by the AVHRR sensor with a spatial resolution of 1.1 km × 1.1 km at nadir. A first thermal-infrared TIT retrieval study using AVHRR was presented by Groves and Stringer [1991]. They applied Kuhn et al. [1975]'s method as well as a simple theoretical approach (Maykut [1986]). In spite of using realistic AVHRR surface temperatures, the retrieved ice-thickness results were ambiguous. They postulated that this may result from simplified parameterizations of the fluxes within the model or inappropriate atmospheric data. Yu and Rothrock [1996] further improved the retrieval of thin-ice thickness by using AVHRR data in a thermodynamic ice-growth model (Maykut and Untersteiner [1971]) to retrieve the ice thickness up to 50 cm with a relative error of ±20 %. Compared to the previous algorithms this model is more detailed in terms of the calculation of the net total radiation balance, the separate treatment of the turbulent heat fluxes and the conductivity of the ice and snow layer. Based on Yu and Rothrock [1996]'s method Yu et al. [2001], Drucker and Martin [2003], Yu and Lindsay [2003] and Willmes et al. [2010] performed case studies with a slightly modified algorithm for Arctic shelf seas.

Error analyses of thin-ice thickness case studies demonstrated the high agreement between TIT retrieval results and other TIT data sets (e.g., about 80 % agreement with moored or submarine upward looking sonar; Drucker and Martin [2003], Wang et al. [2010]).

A detailed description of the thin-ice thickness retrieval as it is used in this thesis is provided in Section 4.2.

2.2.3 Thermal-infrared sea-ice concentration

The following approaches to calculate sea-ice concentration (SIC) are based on the assumption that the ice-surface temperature includes information about the fraction of open water within a pixel (Drüe and Heinemann [2004]).

Ishikawa et al. [1996] calculated SIC using AVHRR T_B. They used the brightness temperature assuming that the physical and brightness temperature are identical. According to Tanaka et al. [1985], this assumption is reasonable for polar regions because of low water vapor contents in the atmosphere. For each AVHRR scene, they defined a temperature for thick ice outside the polynya to distinguish between polynya and drift/fast ice. This temperature is referred to as background temperature T_{bg}. The freezing temperature of open water T_f is set to -1.8 °C. SIC is calculated as follows:

$$SIC = 100 \times ((T_B - T_f)/(T_{bg} - T_f)) \qquad (2)$$

Polynya area (POLA) is calculated using all pixels with a SIC lower than 50 %.

Ciappa et al. [2012] modified this algorithm to determine POLA in the Terra Nova Bay polynya (Antarctic). They omitted the intermediate step to calculate SIC, but iteratively calculated a temperature threshold T_{th} to bound the POLA:

$$T_{th} = (T_{max} + T_{bg})/2 \qquad (3)$$

All pixels classified warmer than T_{th} were rated as part of the polynya.

Instead of a constant T_f the authors used the maximum temperature T_{max} in a MODIS scene as open-water temperature. The initial background temperature T_{bg} was specified 1 °C lower than T_{max}. During the further calculation process T_{bg} was determined using the pixels of a strip next to the boundary of the polynya. Iteratively adjusting T_{bg} and T_{th}, POLA was enlarged until it converges with T_{bg}.

Drüe and Heinemann [2004] calculated the SIC with MODIS ice-surface temperatures T_s. Their potential open-water algorithm (POTOWA) was defined as follows:

$$SIC = 100\%; \qquad T_s \leq T_{bg}$$
$$SIC = \left(1 - \left((T_s - T_{bg})/(T_f - T_{bg})\right)\right) \times 100; \quad T_{bg} < T_s < T_f \qquad (4)$$
$$SIC = 0\%; \qquad T_s \geq T_f$$

with a constant T_f of -1.8 °C. T_{bg} was determined within a 50 × 50 pixel kernel using a bilinear function. The obtained high-resolution SIC product has a spatial resolution of 1 km × 1 km. The assessment study of Drüe and Heinemann [2004] stated that the error of the sea-ice concentrations is approximately ±10 %. Willmes et al. [2010] applied the method to the Laptev Sea and showed that the estimated polynya area is in agreement with other data sets.

2.2.4 Passive microwave sea-ice concentration

A variety of sea-ice concentration (SIC) products are based on calculations from passive microwave data. SSM/I brightness temperature at 19, 37 and 85 GHz has been used to retrieve SIC products with a gridded spatial resolution of 25 km × 25 km for the last three decades (Cavalieri et al. [1996], Meier et al. [2006]). Several retrievals based on SSM/I data (e.g., Bootstrap, NASA TEAM) were compared by Andersen et al. [2007]. The algorithms were adjusted and enhanced for the higher spatial resolution of AMSR-E data (Markus et al. [2008], Markus et al. [2011]).

In this thesis, sea-ice concentrations are used that are derived through the ARTIST sea ice (ASI) algorithm (Spreen et al. [2005], Spreen et al. [2008]). The algorithm is based on the polarization difference P_d between the vertically and horizontally polarized AMSR-E 89 GHz T_B. Two tie points representing the P_d of open water (SIC = 0 %) and the P_d of consolidated ice (SIC = 100 %) are required to derive the sea-ice concentration. The choice of the tie points is important for the accuracy of the SIC retrieval. The AMSR-E 89 GHz channels are sensitive to water vapor and clouds; therefore the algorithm includes the opacity of the atmosphere. The opacity defines the transmission factor of the atmosphere. Sea-ice concentrations for 0 % and 100 % are calculated as follows:

$$SIC = \left(\frac{P_d}{P_0} - 1\right)\left(\frac{P_{s,w}}{P_{s,i} - P_{s,w}}\right) \qquad \text{for } SIC \to 0\% \qquad (5)$$

$$SIC = \left(\frac{P_d}{P_1} + \left(\frac{P_d}{P_1} - 1\right)\left(\frac{P_{s,w}}{P_{s,i} - P_{s,w}}\right)\right) \times 100 \quad \text{for } SIC \to 100\% \quad (6)$$

where P_d is the polarization difference, P_0 and P_1 are the polarization differences including the atmospheric influence for SIC = 0 % (open water) and SIC = 100 % (closed ice cover) and $P_{s,w}$ and $P_{s,i}$ are the polarization differences for water and ice, respectively. The sea-ice concentrations between 0 % and 100 % are interpolated with a third order polynomial:

$$SIC = \left(d_3 P_d^3 + d_2 P_d^2 + d_1 P_d + d_0\right) \times 100 \quad (7)$$

where d_0-d_3 are determined with Equation 5 and quation 6 and their first derivatives by solving a linear equation system.

Spreen et al. [2008] added a weather filter to the ASI algorithm to reduce effects of water-vapor and clouds, which disturb the microwave signal that is emitted from the earth's surface. Therefore, Spreen et al. [2008] used the lower non-weather influenced frequencies at 18.7, 23.8 and 36.5 GHz. After the atmospheric correction, only heavy rain events result in overestimated ice concentrations over the open ocean (Spreen et al. [2008], Kaleschke [2003]). The final sea-ice concentration product has a spatial resolution of 6.25 km × 6.25 km and this data is available for the last 10 years.

According to Spreen et al. [2008], the ASI algorithm yields good results for high SIC. For ice concentrations above 65 % the error can be up to 10 %; in regions of thin ice the error can be even higher (Kwok et al. [2007], Andersen et al. [2007]). In these regions (e.g., polynyas) the passive microwave sea-ice concentrations are underestimated. That can be explained by the similar microwave emissivity of very thin ice and open water. With increasing ice thickness the emissivity gradually increases to a level similar to first-year ice (Comiso and Steffen [2001]). Hence, the similarity of thin-ice and open-water emissivity has a significant influence on the ice-concentration retrieval if a large part of the FOV is covered by thin ice ([Kwok et al. [2007]). This means that ice-concentration maps do not follow the strict definition of sea-ice concentration (100 % SIC if the pixel is fully covered by ice) in regions of thin ice (Comiso and Steffen [2001], Kwok et al. [2007]). However, according to Comiso and Steffen, [2001] the underestimation of SIC increases the value of the ice-concentration maps when thin-ice regions within polynyas are monitored because they can be distinguished from fast and drift ice.

Because of AMSR-E's finer spatial resolution (6.25 km × 6.25 km) the error due to mixed pixels near the coast is reduced in comparison to SSM/I SIC. This improves the monitoring of near-coast polynyas.

2.2.5 Passive microwave thin-ice thickness

Passive microwave brightness temperature increases with ice thickness up to approximately 20 cm (Naoki et al. [2008]). As stated by Naoki et al. [2008], the near-surface brine distribution is a key factor that contributes to this relationship because its amount changes within the ice layer when the ice grows. Changes in near-surface salinity result in the modification of the ice layer's dielectric properties affecting the passive microwave emissivity. Thus, the dependence of dielectric properties on brine gives information about the thin-ice thickness (Naoki et al. [2008], Ukita et al. [2000]).

Several thin-ice thickness retrievals based on passive microwave data were developed in the last decade. For these methods, polarization ratios of the SSM/I 37 and 85 GHz channels and AMSR-E 36 and 89 GHz channels, respectively, are used. The usage of the polarization ratio instead of T_B reduces the dependence on the ice type and the surface temperature and facilitates the distinction of shelf ice from open water, and hence smaller coastal polynyas (Markus and Burns [1995]). The lower (36, 37 GHz) frequencies are less sensitive to atmospheric influence (mainly water vapor) but have a coarser spatial resolution than the higher (85, 89 GHz) frequency channels.

The inversion of the polarization ratio to ice thickness requires a regression with thermal-infrared thin-ice thickness (see Section 2.2.2). Table 3 summarizes the methods and shows the different regression models to infer ice thickness.

Martin [2004] developed the simple polarization ratio (R_{37}) which is derived using the vertical and horizontal polarized SSM/I 37 GHz channels and applied this approach to the Chukchi Sea (Table 3). The simple polarization ratio (R_{37}) is calculated as follows:

$$R_{37} = \frac{T_{B37V}}{T_{B37H}} \tag{8}$$

The scatter plot in Figure 3 shows that the R_{37} values decrease with increasing AVHRR TIT. At a R_{37} value of 1.1 and an AVHRR ice thickness of approximately 20 cm, a vertical asymptote of R_{37} is approached Martin [2004]. Using an exponential fitting equation, ice thickness below 20 cm are calculated with a gridded spatial resolution of 25 km × 25 km. This method is transferred to the AMSR-E 36 GHz channels (12.5 km × 12.5 km) utilizing the finer spatial resolution of the AMSR-E sensor (Martin [2005]). The finer spatial resolution allows a more accurate monitoring of smaller polynyas.

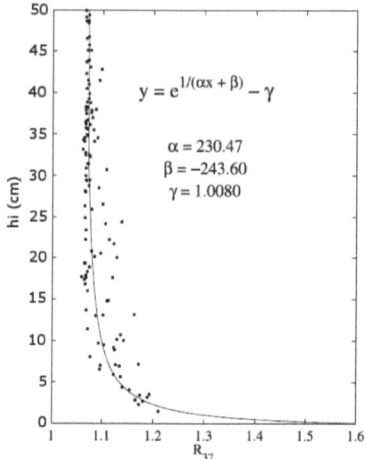

Figure 3: The scatter plot shows the AVHRR thin-ice thickness (interpolated on the 25 km SSM/I grid) on the vertical axis and the SSM/I 37 GHz polarization ratio (R_{37}) on the horizontal axis. The case study is for the Chukchi Sea on 12 March, 2000. © Figure: Martin [2004], Figure 3; modified.

Tamura et al. [2007]'s thin-ice thickness algorithm is based on the polarization ratios of the SSM/I passive microwave data from the 85 GHz and 37 GHz channels:

$$PR_{85(37)} = \frac{T_{B85(37)V} - T_{B85(37)H}}{T_{B85(37)V} + T_{B85(37)H}} \qquad (9)$$

This algorithm was developed for polynyas in the Antarctic. After a comparison between AVHRR TIT and PR85 (PR37), linear fitting equations are used to calculate the ice thickness below 20 cm (Table 3). This algorithm utilizes the higher spatial resolution of the 85 GHz channel for the retrieval of very thin ice (≤ 10 cm). Tamura et al. [2007] suggested that the usage of the coarser resolution 37 GHz channel, which is less sensitive to atmospheric conditions, is sufficient for the retrieval of ice thickness between 10 and 20 cm. As mentioned above, the 85 GHz channel is influenced by water vapor and clouds resulting in an overestimation of the ice thickness. Thus, a correction of the atmospheric influence is applied using the 37 GHz channel to eliminate contaminated pixels. Using these TIT results, Tamura et al. [2008] calculated the ice production in Antarctic polynyas for the last 20 years. Based on these two studies, Tamura and Oshima [2011] made use of a slightly modified algorithm to determine the thin-ice thickness distribution within Arctic polynyas. Instead of two linear fitting equations, Tamura and Oshima [2011] used four equations (two linear and two exponential fitting curves) to calculate the ice thickness within four ice-thickness classes with a maximum ice thickness of 15 cm (Figure 4; Table 3). The ice-thickness results were then used to calculate the ice production in Arctic polynyas.

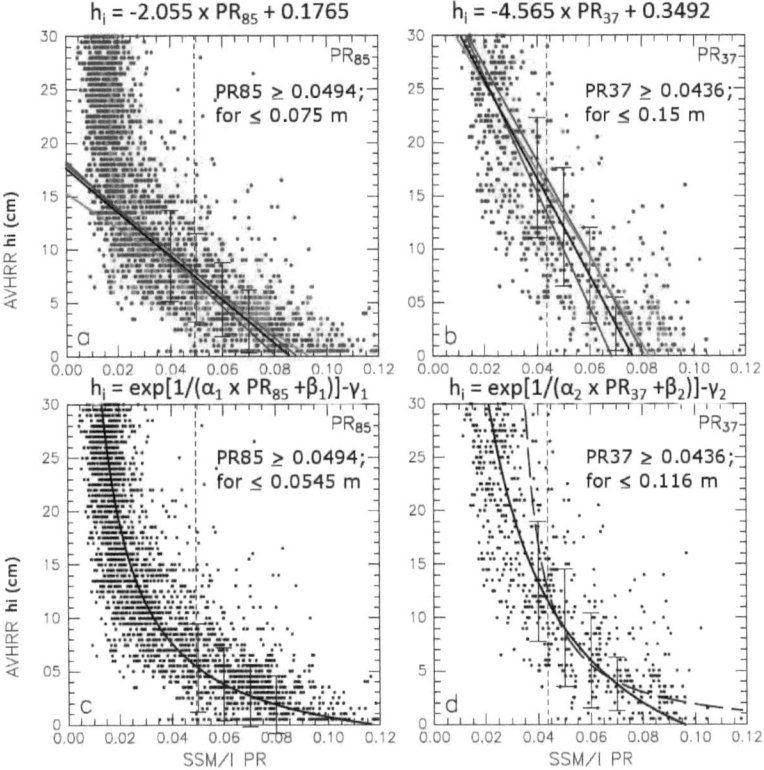

Figure 4: The scatter plots show the AVHRR thin-ice thickness on the vertical axis and the SSM/I polarization ratio (PR) for (a, c) 85 GHz and (b, d) 37 GHz on the horizontal axis. The solid black lines in (a) and (b) denote the linear lines and in (c) and (d) the exponential curves of the equations written above the subplots. The vertical lines with crossbars show the standard deviation of the plots with respect to the linear line and the exponential curve. The dashed vertical lines show the PR thresholds. The blue, red and green lines in (a) and (b) are the principal component axes calculated for the North-Water polynya, the Chukchi polynya and the Laptev polynya. The blue, red and green points are from the North-Water polynya, the Chukchi polynya and the Laptev polynya. The fitting equation shown by the solid curve in (d) is close to that of Martin [2004] shown by the dashed line in (d). © Figure: Tamura and Oshima [2011], Figure 3; modified.

Table 3: Overview of methods used to derive thin-ice thickness (TIT) within polynyas from polarization ratios of passive microwave data and a regression with thermal thin-ice data. The symbols are explained in the following: R and PR are polarization ratios; T_B is the brightness temperature; h_i is the ice thickness; α, β and γ are regression coefficients; the indices 36, 37, 85, 89 refer to the used frequency; the indices V and H refer to vertical and horizontal polarization, respectively.

Reference	Sensor Frequency	Polarization ratio	Regression model	Spatial resolution (km × km)	Region
Martin [2004]	SSM/I 37 GHz	$R_{37}=T_{B,37V}/T_{B,37H}$	$h_{i,37}=\exp[1/(\alpha R_{37}+\beta)]-\gamma$ α=230.5, β=243.6, γ=1.008	25 × 25	Chukchi Sea
Martin [2005]	AMSR-E 36 GHz	$R_{36}=T_{B,36V}/T_{B,36H}$ $R_{36,cal}=(R_{36}-b)/a$ a= 1.3264, b= 0.3444	$h_{i,36}=\exp[1/(\alpha R_{36,cal}+\beta)]-\gamma$ α=230.5, β=243.6, γ=1.008	12.5 × 12.5	Chukchi Sea
Tamura et al. [2007]	SSM/I 37, 85 GHz	$PR_{85}=$ $(T_{B85,V}-T_{B,85H})/(T_{B,85V}+T_{B,85H})$ $PR_{37}=$ $(T_{B,37V}-T_{B,37H})/(T_{B,37V}+T_{B,37H})$	$h_{i,85}=-3.912\times PR_{85}+0.301$ → $PR_{85} \geq 0.0495$; $h_i = 0\text{-}10$ cm $h_{i,37}=-9.02\times PR_{37}+0.7125$ → $PR_{37} \geq 0.0571$; $h_i = 0\text{-}20$ cm	12.5 × 12.5	Antarctic Ocean
Willmes et al. [2010]	SSM/I 85 GHz	$R_{85}=T_{B,85V}/T_{B,85H}$	$h_{i,85}=\exp(5.2\times R_{85})\times 0.0002$	12.5 × 12.5	Laptev Sea
Willmes et al. [2010]	AMSR-E 89 GHz	$R_{89}=T_{B,89V}/T_{B,89H}$	$h_{i,89}=\exp(-6.2\times R_{89})\times 86.2$	6.25 × 6.25	Laptev Sea
Willmes et al. [2010]	AMSR-E 36 GHz	$R_{36}=T_{B,36V}/T_{B,36H}$	$h_{i,36}=\exp(2.8\times R_{36})\times 0.002$	12.5 × 12.5	Laptev Sea
Willmes et al. [2010]	AMSR-E SIR 36 GHz	$R_{36sir}=T_{B,36Vsir}/T_{B,36Hsir}$	$h_{i,36sir}=\exp(-5.49\times R_{36sir})\times 48.59$	7.5 × 7.5	Laptev Sea
Tamura and Oshima [2011]	SSM/I 37, 85 GHz	$PR_{85}=$ $(T_{B,85V}-T_{B,85H})/(T_{B,85V}+T_{B,85H})$ $PR_{37}=$ $(T_{B,37V}-T_{B,37H})/(T_{B,37V}+T_{B,37H})$	$h_{i,85}=-2.055\times PR_{85}+0.1765$ → $PR_{85} \geq 0.0494$; $h_i \leq 0.75$ cm $h_{i,37}=-4.565\times PR_{37}+0.3492$ → $PR_{37} \geq 0.0436$; $h_i \leq 15$ cm $h_{i,85}=\exp[1/(\alpha_1\times PR_{85}+\beta_1)]-\gamma_1$ → $PR_{85} \geq 0.0494$; $h_i \leq 0.545$ cm $h_{i,37}=\exp[1/(\alpha_2\times PR_{37}+\beta_2)]-\gamma_2$ → $PR_{37} \geq 0.0436$; $h_i \leq 11.6$ cm $\alpha_1=215.15$, $\beta_1=0.508$, $\gamma_1=1.0395$, $\alpha_2=88.49$, $\beta_2=1.023$, $\gamma_2=1.1113$	12.5 × 12.5	Arctic Ocean

Willmes et al. [2010] took these methods and applied them with slightly changed parameterizations to the Laptev Sea polynya (Table 3). They used SSM/I 85 GHz T_B, AMSR-89 GHz T_B, AMSR-E 36 GHz T_B and AMSR-E enhanced resolution T_B for the calculation of polarization ratios. AMSR-E enhanced resolution data is reprocessed with the Scatterometer Image Reconstruction (SIR) method (AMSR-E SIR) and has a spatial resolution of 7.5 km × 7.5 km (Early and Long [2001]).

To infer TIT, exponential regression models between the polarization ratios and AVHRR thermal-infrared thin-ice thickness were applied. Willmes et al. [2010] stated that the TIT calculated using the AMSR-E SIR 36 GHz data (TIT_{36sir}) is valuable for the operational retrieval of ice thickness below 20 cm in the Laptev Sea due to the data's sufficient spatial resolution and the negligible atmospheric disturbance on the 36 GHz channel. When TIT, that is calculated based on the AMSR-E 89 GHz channel (TIT_{89}), is used and applied to longer time periods, one has to be careful because of the atmospheric interference on this channel. Altogether, the study showed that passive microwave data is applicable for the TIT retrieval within the Laptev Sea polynya. However, the spatial resolution is a key factor and highly influences the quality of the results. For narrow polynyas, as occur often in the Laptev Sea, the low polynya area to edge ratio leads to mixed pixels of open water/thin ice and fast/drift ice when coarser resolution passive microwave data is used.

2.2.6 Polynya signature simulation method

The polynya signature simulation method (PSSM) was developed by Markus and Burns [1995] to identify polynya area (POLA) from passive microwave data. The method iteratively classifies open water, thin ice and thick ice using SSM/I 37.5 GHz and 85 GHz T_B. A combination of open-water and thin-ice area describes the polynya area. As in Tamura et al. [2007], the high spatial resolution of the 85 GHz channel and the weak influence of atmospheric constituents on the 37 GHz channel are utilized.

As a first step, the polarization ratio (PR) of the vertically and horizontally polarized 85 GHz frequency is calculated. By applying a predefined thin-ice threshold to the 85 GHz polarization ratio, an initial polynya area is determined. Synthetic 37 GHz images are simulated using the initial polynya-area image whereas sea-ice and open-water pixels are allocated with an average brightness temperature. The two simulated images are convolved with the SSM/I antenna pattern. The measured and the synthetic 37 GHz polarization ratios are compared with each other defining two thresholds for the upper and lower boundary of the polynya area using the maximum correlation coefficient and the minimum absolute difference. The three steps of classification, simulation and comparison are repeated until the best fit between the measured and simulated 37 GHz images is achieved (Markus and Burns [1995]).

In previous studies, different thin-ice thresholds for PR were applied for different research areas. For instance, Kern et al. [2007] applied a thin-ice threshold of 0.085 to derive polynya area in the Ross Sea (Antarctic).

Willmes et al. [2010] applied PSSM to AMSR-E 36 GHz and 89 GHz T_B and adjusted the thin-ice threshold for PR_{89} to 0.07 for the Laptev Sea to get the best fit of polynya area with visible and Advanced Synthetic Aperture Radar (ASAR) data. They stated that in the Laptev Sea the PSSM polynya area includes ice thickness below 20 cm.

2.2.7 Methods used in this thesis

In this thesis, several methods are applied to derive polynya characteristics, like polynya area or thin-ice thickness distribution. The methods are:
- the thermal-infrared thin-ice thickness retrieval (Section 2.2.2);
- the passive microwave thin-ice thickness retrievals (Section 2.2.5);
- the polynya signature simulation method (PSSM, Section 2.2.6).

The work is focused on the thermal-infrared thin-ice thickness retrieval. Section 4 deals with the improvement of the algorithm and provides a thorough sensitivity analysis of the thin-ice thickness results.

For the different retrievals of polynya characteristics the following data products are used:
- MODIS ice-surface temperature;
- MODIS sea-ice concentration;
- AMSR-E brightness temperature;
- AMSR-E sea-ice concentration.

These data products with their specifications are described in the subsequent section.

3 Data sets

Three types of data sets are used in this thesis (Figure 5). The main focus lies on satellite remote sensing data (optical and passive microwave) available with high spatial coverage as well as fine spatial and temporal resolution. These data sets are used to derive the following polynya variables: thin-ice thickness, sea-ice concentration and polynya area. In-situ data measured during the Transdrift XV expedition in the Siberian Arctic in 2009 is used as a verification data set for the optical remote sensing data. As a third data type, model data sets are used. Atmospheric model data supports the retrieval of thin-ice thickness from satellite data. Simulated sea-ice concentration from a coupled sea-ice/ocean model is evaluated by remote sensing data to show the ability of this model to simulate polynyas.

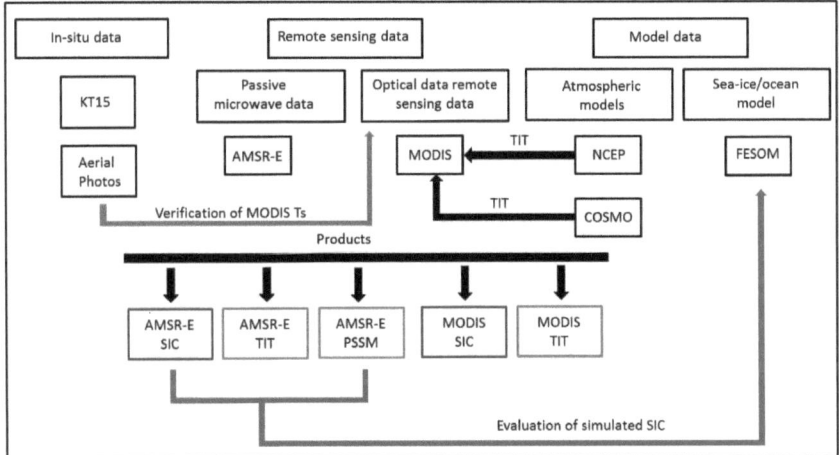

Figure 5: Overview of the data sets and their relationship to each other. Red boxes indicate that products are acquired from others (see Section 3.1); green boxes indicate that products are derived by the author using the methods mentioned in Section 2.2.

3.1 Satellite remote sensing data sets

The following satellite remote sensing data sets are used:
- optical remote sensing data measured by the Moderate-resolution Imaging Spectroradiometer (MODIS);
- passive microwave data detected by the Advanced Microwave Scanning Radiometer – Earth Observation System (AMSR-E);
- backscatter coefficients measured by Environmental Satellite (Envisat) Advanced Synthetic Aperture Radar (ASAR).

These data sets provide data measured within various portions of the electromagnetic spectrum, each with different spatial resolution and coverage. How these different measurement characteristics influence the thin-ice thickness monitoring will be discussed in Sections 4.

3.1.1 Optical remote sensing data

In this thesis, the MODIS Terra and Aqua Sea Ice Extent 5-min L2 Swath product is used (MOD/MYD29; Hall et al. [2007]). Sea ice-surface temperature required for the retrieval of thin-ice thickness within the polynya is stored in the MOD/MYD29 product.

The ice-surface temperature is automatically generated using the following MODIS products: MOD/MYD021KM Level 1B calibrated radiances (Guenther et al. [2002]), MOD/MYD03 geolocation (Wolfe et al. [2002]) and MOD/MYD35_L2 cloud mask (Riggs et al. [2012], Ackerman et al. [1998]). The data is provided with a spatial resolution of 1 km × 1 km at nadir and an accuracy of ±1.6 °C (Hall et al. [2004]). The nominal swath coverage is 2330 km (cross track) × 2030 km (along track). The data product is available through the US National Aeronautics and Space Administration's (NASA) Next Generation Earth Science Discovery Tool (http://reverb.echo.nasa.gov/reverb).

Because the retrieval of reliable ice-surface temperatures is only applicable during clear sky conditions, the quality-controlled MODIS cloud mask (MOD/MYD35_L2) is applied to the MODIS T_s product to exclude cloudy regions (Ackerman et al. [1998], Hall et al. [2004], Frey et al. [2008]). The cloud-mask algorithm has difficulties in identifying sea smoke and thin low clouds resulting in overestimated high ice-surface temperatures where these clouds occur (Ackerman et al. [1998]).

The MODIS cloud mask is derived using all available MODIS data. The data measured with the visible channels is not usable at night for the cloud-mask retrieval. Thus, the quality of the cloud mask for MODIS night scenes is lower than for day images. Recently, the thin-ice thickness calculation is only applicable for MODIS night scenes. Therefore, the presence of clouds that are not identified by the cloud mask may occur more often.

MODIS sea-ice concentrations are calculated according to Drüe and Heinemann [2004]. The daily data set has a spatial resolution of 1 km × 1 km and covers the Laptev Sea region. The data is provided by the S. Willmes, University of Trier (personal communication, 2011). MODIS SIC are used during the computing process of daily MODIS thin-ice thickness composites as a comparison and correction data set.

3.1.2 Passive microwave data

AMSR-E / Aqua Daily L3 brightness temperatures (AMSR-E T_B) are used for the retrieval of thin-ice thickness and for the polynya signature simulation method. The data is acquired by the US National Snow and Ice Data Center (NSIDC) (Cavalieri et al. [2004]). For the retrieval methods, AMSR-E T_B at 36.5 GHz with a gridded spatial resolution of 12.5 km × 12.5 km and at 89 GHz with a gridded spatial resolution of 6.25 km × 6.25 km are required.

AMSR-E sea-ice concentrations calculated by the ASI algorithm are used for the derivation of open-water and polynya area. The SIC product with a spatial resolution of 6.25 km × 6.25 km is provided by the University of Hamburg (Kaleschke et al. [2001], Spreen et al. [2008]).

AMSR-E enhanced resolution data is generated from irregularly sampled data using the Scatterometer Image Reconstruction (SIR) algorithm (Early and Long [2001]). This data set is referred to as AMSR-E SIR. The spatial resolution of the product is 7.5 km × 7.5 km. The AMSR-E SIR data is obtained by the NSIDC (Long and Stroeve [2011]).

3.1.3 Envisat ASAR

Envisat ASAR wide swath backscatter data is used for comparison with other remote sensing data sets. Polynya edges can be defined by manual interpretation of this high-resolution data set. The instrument operates at C-band (5.34 GHz) with vertical co-polarization (VV) and covers approximately 400 km × 800 km with a spatial resolution of 150 m × 150 m. ASAR Level 1 data is provided by the European Space Agency (ESA).

3.2 In-situ data sets

The thesis is a component of the German-Russian project 'System Laptev Sea – Investigation of polynyas and front systems'. The aims of the project are (1) to describe the annual and spatial variability of oceanographic fronts and transport processes, (2) to investigate the reaction of polynya systems to changing forcing variables, (3) to examine the interaction between sea floor, polynya and sea ice in terms of microbiology and biogeochemistry and (4) to monitor the spatio-temporal variability of the polynya dynamics and the exchange between atmosphere, ocean and sea ice within the polynya system.

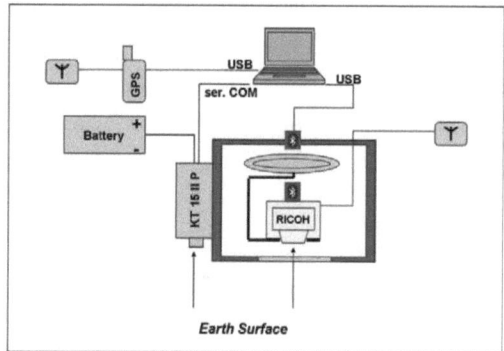

Figure 6: Measurement system for remote sensing data (ice-surface temperature and aerial photographs) in the helicopter. The green box is fixed in the helicopter's hatch. The aerial photographs are taken with the camera fixed within the box. The ice-surface temperature is measured with the KT 15 II P pyrometer outside of the box. In addition to the camera's 1Hz-GPS, an external 1Hz-GPS is used for tracking the flights. © Figure: A. Helbig, University of Trier (2009).

The Transdrift XV expedition, which took place in the Siberian Arctic during March and April, 2009 was part of this project. Measurements in the research field of oceanography, meteorology and remote sensing were acquired. For oceanographic purposes, moorings equipped with Acoustic Doppler Current Profiler (ADCP) and Conductivity Temperature Depth Sensor (CDT) were deployed near the fast-ice edge during the expedition. Additionally, salinity, temperature, oxygen and chlorophyll were measured during the

expedition's ice camps. Two automatic weather stations (AWS) were installed near the fast-ice edge to document meteorological conditions during the expedition period (Figure 7). The AWS measured air temperature, humidity, net radiation, wind speed and wind direction.
For remote sensing purposes a measurement system was installed in the helicopter to measure ice-surface temperatures and to take aerial photographs of the polynya conditions (Figure 6, Figure 8).

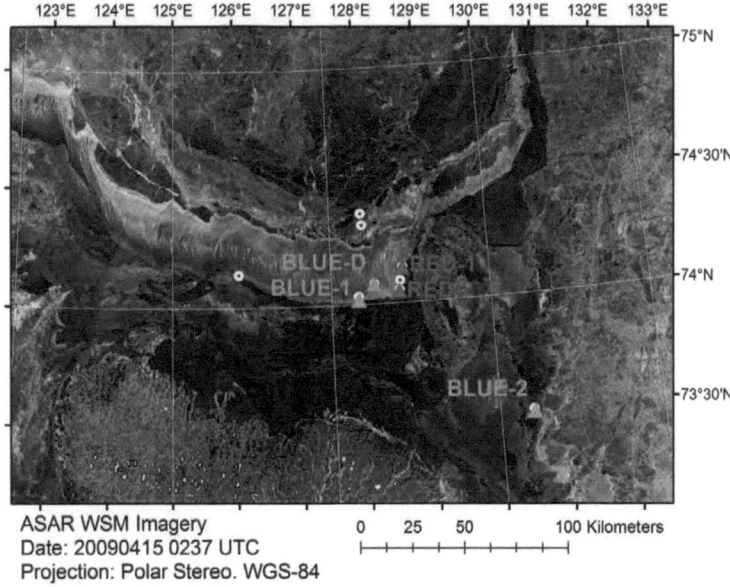

Figure 7: The positions of the automatic weather stations (AWS) during the Transdrift XV expedition are denoted within the map. The AWS BLUE was first installed near the fast-ice edge at the position BLUE-1 and drifted on an ice floe to position BLUE-D. After the drift, the AWS BLUE was reinstalled on the fast ice at the position BLUE-2. The AWS RED was at first installed at the position RED-1. After some repairs, the AWS was reinstalled at the position RED-2. In the background an Envisat ASAR image detected on 15 April, 2009 0237 UTC is shown. The yellow circles denote the position of the ice camps. © Envisat ASAR image: T. Krumpen, AWI (2009).

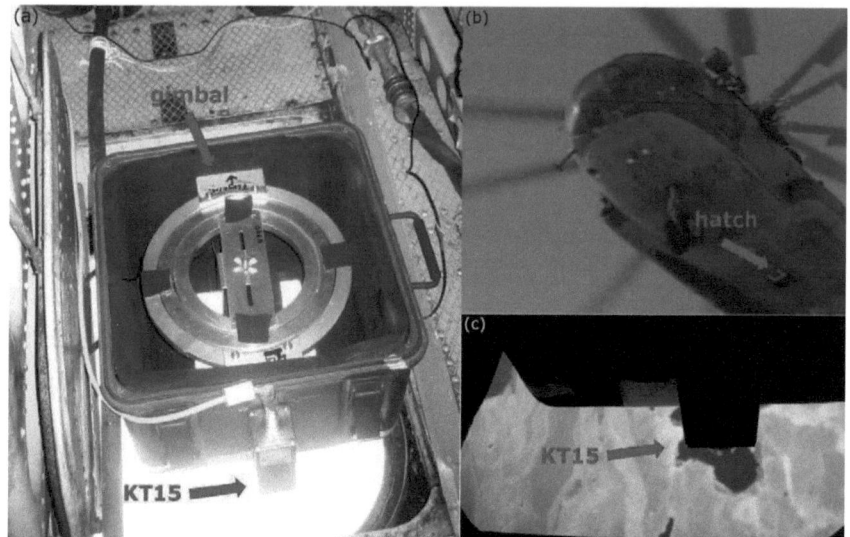

Figure 8: (a) Inside the helicopter: green box is installed in the helicopter's hatch. The digital camera is fixed in the gimbal. The KT15 pyrometer is fixed outside of the box. (b) Box with measurement devices is seen through the open hatch of the helicopter. (c) View through the helicopter's hatch during a flight. © Photographs: A. Helbig, University of Trier (2009).

The KT 15 II pyrometer (KT15) was used to detect across-polynya profiles of ice-surface temperature within the polynya (Heitronics Infrarot Messtechnik GmbH [2012]). The instrument measures the emitted long-wave radiation in the wavelength range from 9.0 to 11.5 µm. Within this wavelength range the influence of atmospheric gases (especially water vapor and CO_2) is very low. The flight altitude is therefore not important. The device provides temperature values for emission coefficients between 0 and 1. For all of the measurements an emission coefficient of 1.0 was used. The device measures in the range of -50 to 200 °C. The mean error is 0.03 °C. A temperature measurement is acquired every second. Given a nominal helicopter flying altitude of 100 m the spatial resolution of the surface-temperature measurements is 4 m × 4 m.

Table 4: Details of the polynya transects measured during the aerial survey.

Date	Time (UTC)	Start station	Length (km)	Number polynya crossings	Air temperature (°C)	Wind speed (m s^{-1})	Wind direction
26 March 2009	0830	TI09-3	85	6	-14	5	SW
27 March 2009	0630	TI09-4	85	6	-15	3	SW
1 April 2009	0330	TI09-5	15	1	-23.3	2	S
8 April 2009	0550	TI09-8	100	4	-9.2	5	SE
14 April 2009	0600	TI09-10	100	4	-18.5	9	S
15 April 2009	0610	TI09-11	150	3	-11.6	13	S
21 April 2009	0500	TI09-13	60	2	-10.8	10	SSW

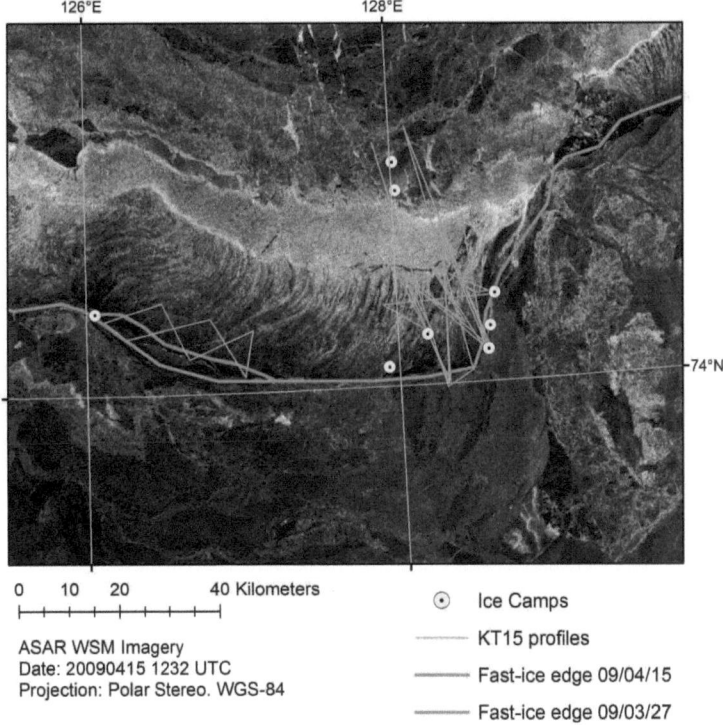

Figure 9: Overview of the KT15 profiles measured during the Transdrift XV expedition. Fast-ice edge at 27 March, 2009 and 15 April, 2009 as well as ice camps are shown. Where the green line for the fast-ice edge in March is not visible it is coincident with the fast-ice edge in April. In the background an Envisat ASAR image detected at 15 April, 2009 1232 UTC is shown. © Envisat ASAR image: T. Krumpen, AWI (2009).

Photogrammetric sea-ice imagery was taken with a RICOH digital camera. The camera was fixed in a gimbal inside the box. At 100 m flight altitude the image dimensions are 120 m × 80 m.

For geolocation, the camera's internal 1Hz-GPS device was used in addition to a similar device. The average flight altitude was 100 m and the average flight speed of the helicopter was 120 km h^{-1}. In total, seven transects were measured between 26 March and 21 April, 2009 (Figure 9). All transects cover the Western New Siberian (WNS) polynya and include several across-polynya legs. Table 4 gives an overview of the length and other details of the polynya transects.

3.3 Model data sets

Simulation outputs of two different model types are used:
- atmospheric models;
- coupled sea-ice/ocean models.

Atmospheric data is required for the retrieval of thermal-infrared thin-ice thickness. In combination with the MODIS ice-surface temperature, the ice thickness is retrieved with a surface energy balance model. For solving the atmospheric flux equations atmospheric variables from these models are used.

Coupled sea-ice ocean models simulate the ice coverage of the ocean. The modeled sea-ice concentration is compared in terms of polynya simulation with sea-ice products derived from remote sensing data.

3.3.1 Atmospheric model data sets

3.3.1.1 NCEP

Atmospheric variables from the US National Centers for Environmental Prediction (NCEP) / Department of Energy (DOE) Reanalysis 2 data set are used to calculate the atmospheric fluxes needed for the thin-ice thickness retrieval (Kanamitsu et al. [2002]). Here, this reanalysis data set is referred to simply as NCEP. The global reanalysis product has a spatial resolution of 1.75° (~200 km grid size in the Laptev Sea area) and a temporal resolution of 6 h. Gridded fields of standard atmospheric variables are available at NOAA's Earth System Research Laboratory. For the purposes of this thesis the following variables are used: 2-m air temperature (T_a), 10-m wind speed (U_{10m}), 2-m specific humidity (q_a) and mean sea level pressure (p) (Table 5).

Table 5: Specifications of the NCEP and COSMO atmospheric data sets. © Table: Adams et al. [2012], Table 1.

Model	NCEP	COSMO
Spatial resolution	1.75° (~200 km) Global	5 km Laptev Sea
Temporal resolution	6 h	1 h
Availability since 2000[1]	2000 – 2012	2002 – 2011
Variables	2-m air temperature, 10-m wind speed, 2-m specific humidity, mean sea level pressure	2-m air temperature, 10-m wind speed, 10-m specific humidity, mean sea level pressure
Quality of data in terms of polynyas	Polynyas not captured resulting in excessively cold 2-m air temperatures above polynyas, warm bias in 2-m air temperature	Polynyas captured using AMSR-E sea-ice fields (sea-ice concentration < 70 % equals to polynya area in model setup); composed data set of daily model runs
Impact on retrieved ice thickness	Due to the frequently occurring warm air temperature bias ice thickness could be overestimated, otherwise air temperatures are underestimated resulting in underestimated ice thickness	Highest accuracy if MODIS and AMSR-E give a similar polynya signal, otherwise underestimation of air temperature resulting in an underestimation of the ice thickness

[1]MODIS ice-surface temperatures available since 2000.

3.3.1.2 COSMO

Consortium for Small-scale modeling (COSMO) is a non-hydrostatic regional weather-prediction model (Steppeler et al. [2003], Schättler et al. [2009]). Since 1999 COSMO has been a main part of the German Weather Service (DWD)'s numerical weather forecast system. Schröder et al., [2011] used the model for the Laptev Sea region and implemented a thermodynamic sea-ice model in COSMO. A high-resolution version of COSMO is nested with two steps in the Global Model Extended (GME with 40 km spatial resolution). AMSR-E SIC is included in the model setup as an initial field of sea-ice conditions. The ice thickness within the polynya is set to 10 cm. Model runs of 30 h (6 h spin-up time inclusive) with daily updated AMSR-E SIC were completed and composed to one data set. The simulations have a horizontal resolution of 5 km and a temporal resolution of 1 h. The data is acquired from D. Schröder (personal communication, 2012). COSMO simulations are abbreviated as COSMO in the following. The variables T_a, U_{10m}, q_a at 10 m and p are used (Table 5).

3.3.2 Sea-ice/ocean model data set

The global coupled sea-ice/ocean model FESOM was developed at the Alfred Wegener Institute for Polar and Marine Research (AWI) in Bremerhaven, Germany. FESOM's sea-ice component is a dynamic-thermodynamic sea-ice model with the thermodynamics following Parkinson and Washington [1979] and an elastic–viscous–plastic rheology according to Hunke and Dukowicz [1997].

To reduce computer costs, the FESOM simulations neglect the horizontal advection (and diffusion) of ocean temperature and salinity (Rollenhagen et al. [2009]). This is appropriate because the dynamics of the Laptev Sea polynyas mainly depend on sea-ice dynamics (e.g., Dethleff et al. [1998]). The ocean model is reduced to the computation of turbulent vertical fluxes of heat and salt as a function of the Richardson number.

Surface stresses between ice and ocean/atmosphere are quadratic functions of the wind speed and the velocity difference respectively. Ocean surface currents and vertical shear required as boundary conditions for sea-ice momentum balance and the ocean vertical mixing scheme have been derived from an annual mean of a fully coupled model run.

Table 6: Specifications of the sea-ice/ocean model data sets. © *Table: Adams et al. [2011], Table 1; modified.*

		FESOM		
		without fast ice, coarse resolution (FESOM-CR)	without fast ice, high resolution (FESOM-HR)	with fast ice, high resolution (FESOM-FI)
Spatial resolution		1/4°	1/20°	1/20°
Temporal resolution		daily mean	daily mean	daily mean
Period		1 Nov 2007 - 11 May 2008	1 April 2008 – 11 May 2008	1 April 2008 – 11 May 2008
Forcing data	Name	NCEP	GME	GME
	Spatial res.	1.875°	0.5°	0.5°
	Temp. res.	daily	6 hourly	6 hourly

This model is used in two different configurations: simulations covering the whole of the Arctic are performed on a rotated 1/4° (approximately 25 km) grid (Rollenhagen et al. [2009]). Starting from a climatological sea-ice distribution, the model is run over several decades forced with a combination of daily NCEP/NCAR reanalysis data for 2-m air temperature and 10-m wind speed, monthly mean humidity from European Centre for Medium-Range Weather Forecasts (ECMWF) reanalysis and climatological fields for precipitation and cloud cover. A horizontal ice volume diffusivity of 2000 $m^2 s^{-1}$ is applied. The time step is 2 h. This coarse-resolution FESOM version is referred to as FESOM-CR.

A second series of simulations is performed with a regional, high-resolution (1/20 °, approximately 5 km) configuration that covers only the Laptev Sea. Starting from an initial sea-ice distribution derived from AMSR-E sea-ice concentrations (daily mean of 1 April 2008) and an initial ice thickness of 1 m with a snow layer of 5 cm, the model is forced with data from the DWD's GME (Majewski et al. [2002]) to specifically simulate the polynya development in April and May, 2008. The horizontal ice volume diffusivity of 100 $m^2 s^{-1}$ is lower than the diffusivity used in FESOM-CR. The time step in this version is 1 h. The fine-resolution FESOM version is abbreviated as FESOM-HR.

Additionally, a third model configuration with a fast-ice mask (FESOM-FI) is used as an enhancement of FESOM-HR. The fast-ice masks are derived monthly from satellite data for the winter season from December to April. The retrieval process of the fast-ice masks is described in detail in Section 4.6.2. According to Lieser [2004], the fast-ice masks are consecutively implemented: the drift velocities within the fast-ice area are set to zero and the sea-ice momentum balance remains unresolved Rozman [2009]. In June, the fast ice is allowed to drift (Bareiss [2003], Bareiss and Görgen [2005]).

The daily sea-ice concentration simulations of the three model configurations are used to prove their ability to simulate polynyas in the Laptev Sea (Table 6). The FESOM-CR sea-ice concentrations cover the entire winter season of 2007/2008. FESOM-HR and FESOM-FI ice concentrations are available for April and May, 2008. FESOM-CR simulations are obtained from R. Timmermann, AWI (personal communication, 2009). FESOM-HR and FESOM-FI simulations are provided by D. Schröder (personal communication, 2009).

4 Retrieval of MODIS thin-ice thickness in the Laptev Sea

This section deals with the retrieval of thermal-infrared thin-ice thickness (TIT) in the Laptev Sea. The retrieved TIT is based on the MODIS ice-surface temperature (T_s) and atmospheric model data. In order to verify MODIS T_s, this data set is compared to in-situ measured high-resolution KT15 T_s (see Section 3.2). After a detailed description of the TIT retrieval, the improvements of the algorithm are compared to the study of Yu and Lindsay [2003]. Furthermore, a sensitivity analysis is accomplished to quantify the uncertainty in the retrieved TIT. In the fifth subsection, the algorithm is applied to two other regions: the Lincoln Sea polynya in the Arctic and the Weddell Sea polynya in the Antarctic. After that, daily MODIS TIT composites are computed and monthly fast-ice masks are derived. Finally, daily MODIS and AMSR-E data sets are discussed with respect to their advantages and drawbacks.

4.1 Verification of MODIS ice-surface temperature

The MODIS T_s product (MOD/MYD29) used here is described by Hall et al. [2004]. They evaluated MODIS T_s separately for the Antarctic and Arctic. In the Antarctic, the MODIS temperature was compared with air temperatures at the South Pole. Including the temperature difference between surface and air, a root mean square error (RMSE) of 1.2 °C was determined. In the Arctic, air temperature data from drifting buoys and a tide station were compared with the MODIS temperatures and a RMSE of 1.3 °C was calculated. Altogether, an accuracy of ±1.6 °C for the MODIS ice-surface temperatures was estimated by Hall et al. [2004].

In this thesis, MODIS T_s is verified for the Laptev Sea using the helicopter-borne, high-resolution (4 m × 4 m) KT15 T_s measurements from the Transdrift XV expedition. The maps in Figure 10 give a comparison of KT15 T_s, MODIS T_s and Envisat ASAR backscatter on 15 April, 2009. The Envisat ASAR image illustrates the polynya conditions with a spatial resolution of 150 m × 150 m. Although the polynya boundaries can be estimated by Envisat ASAR backscatter, it is not possible to derive the ice thickness or ice concentration directly from this image. By visual interpretation it can be suggested that the black band along the fast-ice edge, B1, consists of open water and very thin ice (Figure 10a). The stripes in B2 denote accumulated frazil ice due to Langmuir circulation. B3 shows an area of consolidated thin ice. This interpretation is supported by the aerial photographs which were taken parallel to the KT15 measurements.

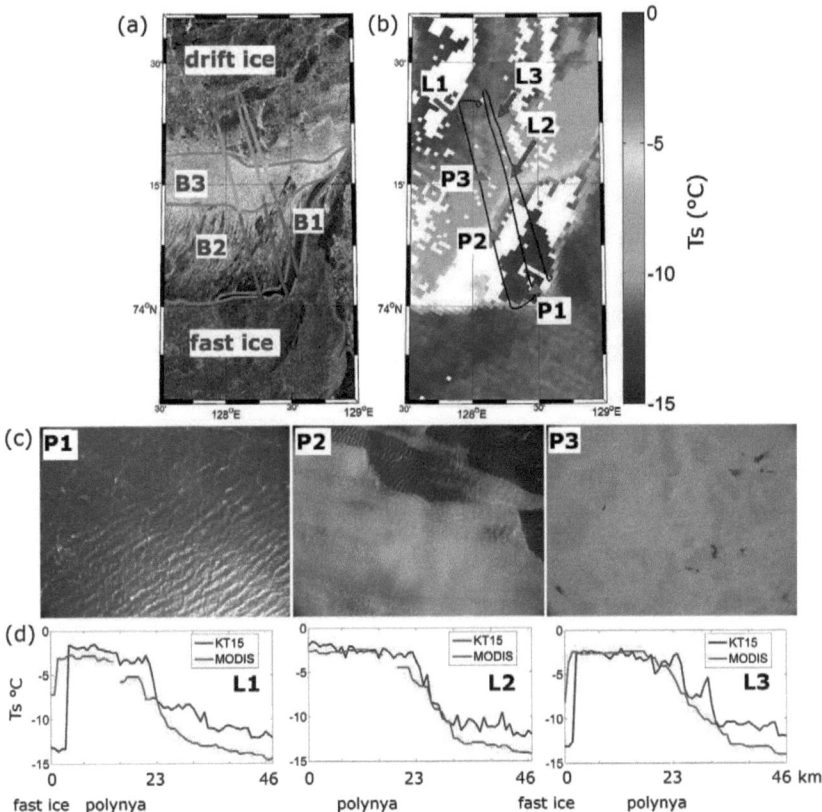

Figure 10: (a) Envisat ASAR image from 15 April, 2009 1232 UTC for illustration of the polynya conditions. Color-graded transect was measured at 15 April, 2009 around 0600 UTC and shows KT15 ice-surface temperature (T_s). Green lines denote the different stages of the polynya: bands (B) 1-3. (b) MODIS T_s from 15 April, 2009 0350 UTC. Data gaps (white areas) result from the MODIS cloud mask. Black line denotes the position of the KT15 transect shown in (a). Gray triangles mark the position of the aerial photographs (P) 1-3. Legs (L) 1-3 denote the profiles shown in (d). (c) Aerial photographs taken parallel to the KT15 data. The picture size is 120 m × 80 m. (d) KT15 T_s across-polynya profiles.

The MODIS and KT15 temperature data sets were measured with a time difference of about 2 h (Figure 10b). The typical across-polynya temperature distribution with high temperatures (near the freezing point of sea water) at the fast-ice edge and decreasing temperatures moving in the direction of the drift ice is shown in both T_s data sets. The strong temperature transition between fast ice (-14 °C) and very thin ice (-1.8 °C) is clearly visible.

The MODIS and the KT15 T_s transects presented in Figure 10d are similar to each other. The temperatures show high agreement in regions of open water and very thin ice. Due to its finer spatial resolution, the KT15 T_s shows a higher variability than the MODIS T_s. In particular in leg L3 between 23 km and 30 km, the temperature peaks are not captured by MODIS T_s. Moreover, a small shift between the MODIS and KT15 T_s jump at the fast-ice edge is seen. This could result from: (1) mixed MODIS pixels including information about fast ice and open water or thin ice and (2) small geolocation errors of both data sets. Two more case studies showing a comparison between KT15 and MODIS data are presented in the Appendix (Figure A1, Figure A2).

Moreover, it becomes obvious that parts of the polynya regions are often covered by clouds or sea smoke. The disturbing influence of these features make the retrieval of T_s and hence TIT from satellite data impossible (Figure 11). Sea smoke appears above the open-water area within polynyas. If cold air passes over the warm ocean, the rapid heating of the air induces convection currents carrying moisture upwards from the ocean surface. The moisture condenses in the cold air and becomes visible as sea smoke. Due to this phenomenon MODIS T_s data is often missing above open-water areas (Figure 10b).

Figure 11: Photograph of the Laptev Sea polynya taken during the Transdrift XV expedition in 2009. Sea smoke above the open-water area of the polynya and the adjacent fast ice is seen.

To get a more objective overview of the relation between KT15 and MODIS T_s, Figure 12 shows a scatter plot of both data sets including four KT15 profiles and corresponding MODIS data with a maximum detection time difference of 2.5 h. The correlation between MODIS and KT15 T_s is high with a correlation coefficient of $r = 0.81$. The outliers could result from: (1) the time difference between the two data sets, (2) mixed MODIS T_s pixels and (3) small geolocation errors of both data sets.

The verification of MODIS T_s with the high-resolution in-situ data supports the accuracy of the satellite data set.

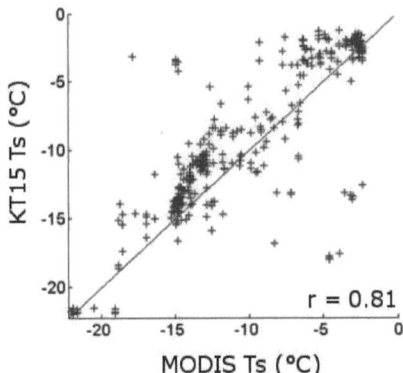

Figure 12: Scatter plot shows KT15 ice-surface temperature (T_s) against MODIS T_s. KT15 profiles of 26 March 2009 0830 UTC, 1 April 2009 0630 UTC, 14 April 2009 0600 UTC, 15 April 2009 0600 UTC and MODIS scenes closest in time (26 March 2009 0905 UTC, 1 April 2009 0400 UTC, 14 April 2009 0445 UTC, 15 April 2009 0350 UTC) to the KT15 profiles are used. The maximum time difference is 2.5 h. The correlation coefficient (r) is 0.81.

4.2 Thin-ice thickness algorithm

The thin-ice thickness (TIT) retrieval based on ice-surface temperatures has been conducted for three decades using an energy balance model approach (see Section 2.2.2). Over time, the algorithm has been continuously improved with respect to more precise parameterizations. The thin-ice thickness retrieval described here is based on the model of Yu and Lindsay [2003]. In order to improve their algorithm, two calculation steps are modified:

(1) the calculation of the turbulent heat fluxes by an iterative stability dependent bulk approach;

(2) improved parameterization of the calculation of the atmospheric emission coefficient required for the determination of the incoming long-wave radiation.

The surface energy balance model is described in Figure 13. The algorithm is based on the condition that the total energy flux to the atmosphere Q_A equals the conductive heat flux through the ice Q_I.

Q_A is calculated as follows:

$$Q_A = Q_0 - H_0 - E_0 \qquad (10)$$

where Q_0 is the net radiation balance, H_0 is the sensible heat flux and E_0 is the latent heat flux. Ice production will occur if the energy flux to the atmosphere Q_A is negative and the water is at its freezing point. The latter is generally observed throughout the winter period.

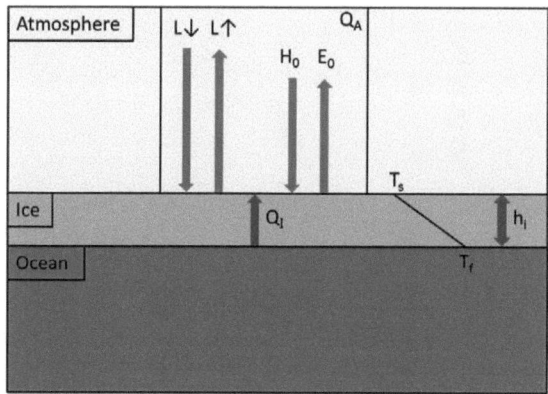

Figure 13: Thin-ice thickness retrieval scheme. L↓ and L↑ are the incoming and outgoing long-wave radiation components, H_0 and E_0 are the turbulent heat fluxes, Q_A is the net energy flux to the atmosphere, Q_I is the conductive heat flux through the ice, T_s is the ice-surface temperature, T_f is the freezing temperature of sea water and h_i is the ice thickness. © Figure: Adams et al. [2012], Figure 2.

Regarding the calculation of Q_0, it has to be mentioned that at this point in time the thin-ice thickness algorithm is only applicable for nighttime MODIS T_s pixels. For an accurate calculation of the short-wave radiation balance, the variables albedo, solar incident radiation and bidirectional reflectance would have to be taken into account (Xiong et al. [2002]). Currently, the parameterization of these variables is difficult due to the lack of adequate data of incident solar radiation and albedo with sufficient spatial and temporal resolution. A pixel-by-pixel calculation of sunrise and sunset is implemented to exclude regions affected by possible incident solar radiation. In this way, daytime conditions are excluded (Wang et al. [2010]), and Q_0 is calculated by the long-wave radiation components only:

$$Q_0 = L\downarrow - L\uparrow \tag{11}$$

The incoming long-wave radiation L↓ is calculated using Stefan-Boltzmann's law:

$$L\downarrow = \varepsilon_a \sigma T_a^4 \tag{12}$$

where T_a is the 2-m air temperature, σ is the Stefan-Boltzmann constant (5.67×10^{-8} J K^{-4} m^{-2} s^{-1}), and ε_a is the atmospheric emission coefficient parameterized according to Jin et al. [2006]:

$$\varepsilon_a = (0.0003(\{T_a\} - 273.16)^2 - 0.0079(\{T_a\} - 273.16) + 1.2983) \times (\{e_a\}/\{T_a\})^{1/7} \tag{13}$$

with T_a in K and e_a as the 2-m water-vapor pressure (in hPa) which is calculated using the specific humidity q_a and the mean sea-level pressure p:

$$e_a = (q_a p)/0.622 \tag{14}$$

The outgoing long-wave radiation L↑ is calculated using T_s and the emission coefficient of the surface ε_s which is assumed to be 1 (Rees [1993]):

$$L\uparrow = \varepsilon_s \sigma T_s^4 \tag{15}$$

The latent heat flux E_0 and the sensible heat flux H_0 are calculated using an iterative bulk approach following Launiainen and Vihma [1990] (Figure 14). This approach includes the atmospheric stability using the Monin-Obukhov similarity theory and the associated universal functions. An advantage of this method is the usage of the input variables wind speed, temperature and humidity at different levels (Launiainen and Vihma [1990]). This allows easily using the NCEP 2-m q_a as well as the COSMO 10-m q_a. Moreover, it allows the choice of an arbitrary calculation level (CL). For the model version used here, CL is at 2-m altitude. Before starting the iteration process the Monin-Obukhov stability is set to 0 (neutral stratification). The potential temperature at the upper level θ_a (here 2 m) and at the lower level θ_s (here surface), the mean potential temperature $\bar{\theta}$, the air density ρ_a, the friction velocity u∗ and the specific humidity at lower level q_s (here surface) are calculated. A constant roughness length for momentum z_0 of 1×10^{-3} m is used. The roughness lengths for temperature z_t and humidity z_q ($z_t = z_q$) are calculated following Andreas [1987]:

$$z_t = f(z_0, \text{Re}) \tag{16}$$

where Re is the roughness Reynolds number. For an average wind speed of 1 m s^{-1}, a z_t of 5.5×10^{-5} m and a C_{HN} (C_H for neutral stratification) of 1.8×10^{-3} is calculated. For stronger winds (U_{10m} = 10 m s^{-1}), a z_t of 2.1×10^{-5} m and a C_{HN} of 1.4×10^{-3} is determined. According to Schröder et al. [2003], a C_{HN} of around 1.5×10^{-3} is appropriate over thin ice. The values calculated using the iterative turbulent flux calculation lie within this range.

The iteration starts with the interpolation of the 10-m wind speed to the 2-m level. After that, the temperature difference between surface and atmosphere ΔT, the friction velocity u∗, the transfer coefficient for heat C_H and evaporation C_E (C_H=C_E) and the difference of specific humidity between surface and atmosphere Δq are determined. Having done this, E_0 and H_0 are calculated:

$$E_0 = \rho_a L C_E \Delta q U_{2m} \tag{17}$$

$$H_0 = \rho_a c_p C_H \Delta T U_{2m} \tag{18}$$

where L is the latent heat of vaporization (2500 J g^{-1}) and c$_p$ is the specific heat of air (1003.5 W s kg^{-1} K^{-1}).

The next step includes the calculation of the Obukhov length L∗ and the universal functions in dependence of the atmospheric stability.

At the end of each iteration (n), the stability is computed:

$$\zeta_n = CL / L_{*,n} \tag{19}$$

The iteration stops if the condition

$$|\zeta_n - \zeta_{n-1}| < 10^{-4} \tag{20}$$

is fulfilled. The convergence criterion is proofed pixel-by-pixel. The iterative approach is designed in a way that a maximum of 25 iterations are allowed. In case of no convergence after 25 iterations, the calculation results for these pixels will be flagged.

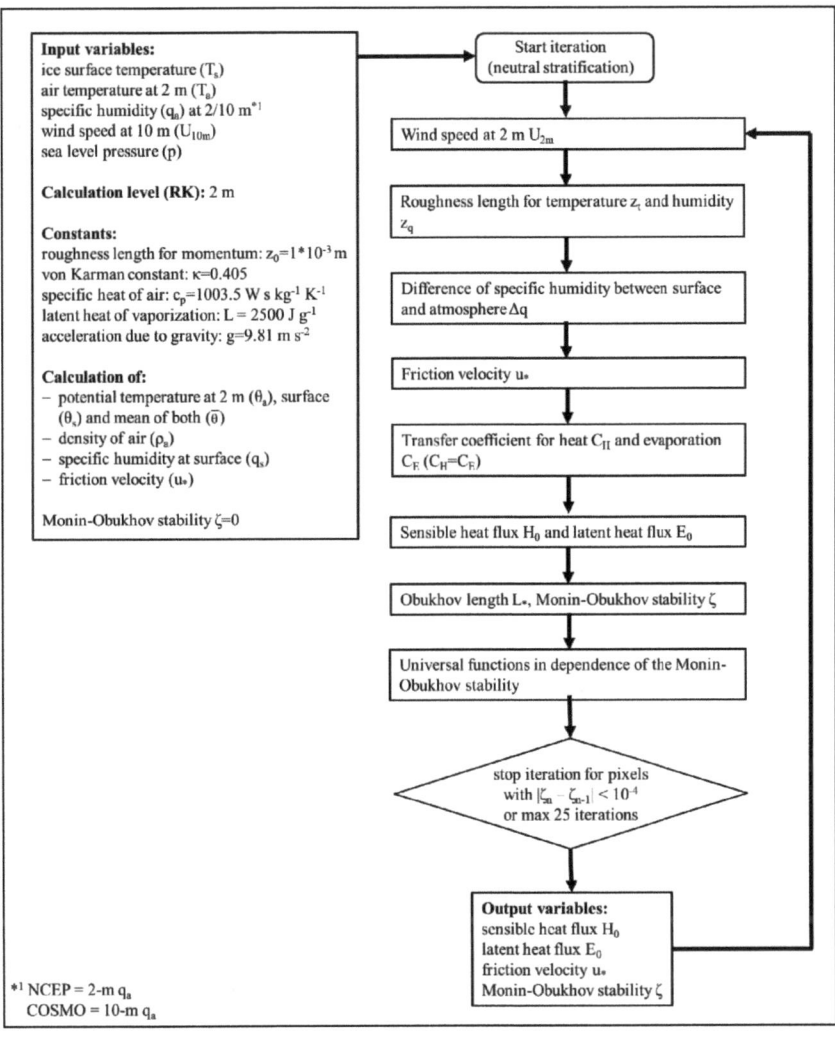

Figure 14: Flow chart of the iterative approach following Launiainen and Vihma [1990] to calculate the turbulent heat fluxes H_0 and E_0 as required for the thin-ice thickness retrieval.

For the calculation of the conductive heat flux through the ice Q_I, a pure ice layer is assumed with a linear temperature profile (Drucker and Martin [2003], Wang et al. [2010]). As in other comparable studies (e.g., Drucker and Martin [2003]) snow on the thin ice is neglected which is in accordance with in-situ observations during the Transdrift XV expedition (see Figure 10, Appendix: Figure A1, Figure A2). The heat flux through the thin-ice layer Q_I can then be calculated as follows:

$$Q_I = k_i(T_s - T_f)/h_i \qquad (21)$$

where k_i is the conductivity of pure ice (2.03 W K^{-1} m^{-1}), T_f is the freezing temperature of sea water and h_i is the ice thickness.

As described above, Q_A equals Q_I which allows solving Equation 21 for the ice thickness h_i:

$$h_i = k_i(T_s - T_f)/Q_A \qquad (22)$$

Although the MODIS cloud mask is applied to the T_s maps, thin clouds and sea smoke are in some cases unidentified, resulting in overestimated T_s. Since polynyas are characterized by warmer surface temperatures similar to the cloud temperature the cloudy pixels appear as polynya pixels. Thus, a thin-ice thickness is falsely calculated for theses pixels. To reduce ice-thickness errors due to the misclassification in the MODIS cloud mask, a polynya mask (LAP) is applied to every single MODIS scene (see Figure 1). The polynya mask broadly covers the regions in the Laptev Sea where polynyas typically occur. The central Laptev Sea is excluded.

As input variables MODIS T_s and atmospheric variables are used to calculate the components of the energy flux to the atmosphere Q_A and hence the TIT. Each atmospheric variable is taken from the two different data sets NCEP and COSMO (see Table 5). These data sets are reprojected to a 1 km × 1 km polar stereographic grid. TIT products are derived using MODIS ice-surface temperatures in combination with:

(1) NCEP atmospheric variables, referred to as TIT$_{MODIS+NCEP}$;

(2) COSMO atmospheric variables, referred to as TIT$_{MODIS+COSMO}$.

Thin-ice thickness distributions are calculated in the Laptev Sea for the two winter seasons 2007/08 and 2008/09 with MODIS T_s and NCEP atmospheric variables. Ice-thickness distributions with MODIS T_s and COSMO atmospheric variables are calculated for case studies in order to compare both products with each other.

4.3 Improvement of the atmospheric flux calculations

In this section, the modifications of the TIT algorithm are analyzed and discussed in comparison to Yu and Lindsay [2003]. $TIT_{MODIS+NCEP}$ up to 50 cm are used. The most significant modification of the TIT algorithm concerns the calculation of the turbulent heat fluxes. Yu and Lindsay [2003] approximated the 2-m wind speed using the geostrophic wind and a reduction factor of 0.34 (Fissel and Tang [1991]). Bulk equations with a constant transfer coefficient for heat (C_H) of 3×10^{-3} for very thin ice and 1.75×10^{-3} for thicker ice are used. Thresholds for very thin ice and thicker ice are not specified in Yu and Lindsay [2003]'s paper.

In contrast, this study applies the iterative approach explained in the previous subsection and takes into account the near-surface atmospheric stratification for C_H. The wind speed is taken from atmospheric model data and adjusted to a height of 2 m using a constant roughness length of 1×10^{-3} m and a variable stability.

Figure 15 shows a histogram of C_H calculated with MODIS and NCEP data for the two winter seasons 2007/08 and 2008/09 for all pixels with an ice thickness between 0 and 50 cm. The maximum frequency of C_H is at 2.0×10^{-3} for both winter seasons. However, larger values of up to 3.8×10^{-3} also occur for very unstable stratification.

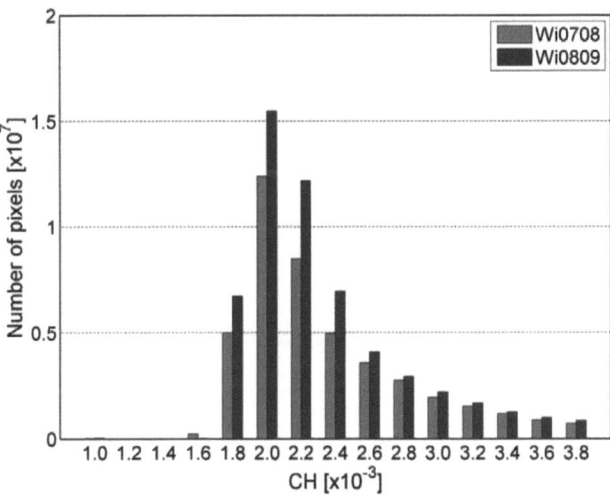

Figure 15: Frequency distribution of the heat transfer coefficient (C_H) (0.2×10^{-3} bins) for the winters 2007/08 (light grey) and 2008/09 (dark grey). The calculation of C_H is based on MODIS ice-surface temperatures and NCEP atmospheric variables. © Figure: Adams et al. [2012], Figure 3.

The constant values for C_H of 1.75×10^{-3} and 3.0×10^{-3} used by Yu and Lindsay [2003] seem to be reasonable compared to the results of this study (using a variable C_H). In order to quantify the effect of these constant transfer coefficients on the TIT retrieval, 10 MODIS scenes with polynya events are selected (Table 7). If a constant C_H of 3×10^{-3} is used instead of a variable coefficient, the ice-thickness values between 0 and 50 cm will be on average 1.9 cm thinner (Table 8). A maximum of 4 cm difference occurs for the 10-20 cm ice class. If a constant C_H of 1.75×10^{-3} is used, the ice thickness will be on average 6.9 cm thinner in comparison to the iterative approach (Table 8).

TIT calculated with the two constant heat transfer coefficients are generally underestimated. For the 10 MODIS scenes, it can be concluded that for simplification purposes a C_H of 3.0×10^{-3} is applicable for TIT up to 50 cm. A constant C_H of 1.75×10^{-3} yields larger ice-thickness differences for all ice classes compared to a variable C_H.

This examination implies that C_H values used in Yu and Lindsay [2003]'s study are suitable as a first approach. However, the iterative approach following Launiainen and Vihma [1990] improves the accuracy of the heat transfer coefficient and thus, the accuracy of the derived TIT.

Table 7: 10 MODIS scenes equally distributed over the two winter seasons 2007/08 and 2008/09. Each scene shows a distinctive polynya event. Some analyses are applied using these scenes as representatives for different polynya situations in the Laptev Sea. © Table: Adams et al. [2012], Table 2.

10 MODIS scenes for analysis purposes		
Satellite	Date	Time (UTC)
Terra	31 December 2007	0520
Terra	27 January 2008	1310
Terra	28 February 2008	1310
Terra	16 April 2008	1310
Terra	23 February 2009	1125
Terra	18 March 2009	0955
Aqua	4 January 2008	0200
Aqua	28 December 2008	2025
Aqua	31 January 2009	0335
Aqua	18 April 2009	1805

Table 8: Comparison of thin-ice thickness (TIT) calculated with a constant heat transfer coefficient (1.75×10-3 and 3.0×10-3) and a variable heat-transfer coefficient using the 10 examples shown in Table 7. Absolute differences between the ice thicknesses for different ice classes are shown. © Table: Adams et al. [2012], Table 3.

TIT classes (cm)	difference C_H 1.75×10^{-3} – variable C_H (cm)	difference C_H 3.0×10^{-3} – variable C_H (cm)
0-10	-3.9	-0.7
10-20	-9.2	-4.0
20-30	-9.9	-3.1
30-40	-8.2	-1.3
40-50	-3.5	-0.7
mean	-6.9	-1.9

A second modification concerns the parameterization scheme of the atmospheric emission coefficient. Yu and Lindsay [2003] applied the parameterization of the atmospheric emission coefficient following Efimova [1961] (EF61). In this study, the atmospheric emission coefficient is determined using a newer improved approach following Jin et al. [2006] (JI06). The main difference between the two approaches is that EF61 parameterizes the emissivity using only the water-vapor pressure and JI06 uses additionally the 2-m air temperature. JI06 improved the parameterization scheme of Brutsaert [1975] for clear-sky polar regions using data from two field campaigns in the Canadian Arctic. The developed empirical model is compared with eleven other parameterization schemes including EF61. JI06's results indicate that their method has the smallest mean error and that their method is applicable to other cold regions.

The calculation of the atmospheric emission coefficient with MODIS T_s and NCEP atmospheric variables using JI06's method for the winter season 2007/08 gives coefficients between 0.59 and 0.82. Using EF61's method the coefficient varies between 0.75 and 0.78, covering a smaller range. Ice thickness is retrieved for the 10 MODIS scenes of Table 7 using JI06's and EF61's parameterizations. A comparison of the TIT distributions shows that JI06's parameterization in general results in thinner ice. The ice thickness is up to 7 cm thinner than using EF61's parameterization scheme. Only in a few cases the ice is thicker using JI06's parameterization with a maximum of 1 cm.

Assuming that JI06's approach gives more accurate results than EF61's approach, the ice thickness in previous studies is mostly slightly overestimated (mean of 10 MODIS scenes between -1.1 and +2.6 cm).

4.4 Sensitivity analysis

The sensitivity analysis is divided into three parts. The first part is based on a statistical sensitivity analysis including the ranking of the input variables required for the TIT retrieval as well as a Monte Carlo error estimation for the determination of an average thin-ice thickness error dependent on the errors of the input variables. The second part includes a comparative sensitivity analysis. Two different atmospheric model data sets are used in combination with MODIS T_s to infer TIT. The two atmospheric data sets differ in their spatial resolution and their handling of polynyas. This analysis is used to identify specific problems and advantages of the atmospheric input variables. As a third sensitivity analysis nearly-coincident MODIS T_s and TIT are used as an uncertainty indicator. A potential error can be derived from their difference assuming that the polynya conditions do not change over a very short time period.

4.4.1 Statistical sensitivity analysis

4.4.1.1 Method

As a first step the importance of each individual input variable on the retrieved TIT in relation to its uncertainty is analyzed. The uncertainties of the input variables T_s, T_a, U_{10m} and relative humidity RH are deduced on the basis of verification studies (Table 9). The following uncertainties based on Renfrew et al. [2002], Hall et al. [2004] and Ernsdorf et al. [2011] are assumed: $\Delta T_s = \pm 1.6$ °C, $\Delta T_a = \pm 4.5$ °C, $\Delta U_{10m} = \pm 1.3$ m s^{-1}, $\Delta RH = \pm 20$ %. The relative humidity is used instead of the specific humidity to avoid inconsistencies between air temperature and saturation humidity. The analysis is performed using the MODIS scenes of Table 7. TIT fields are calculated by varying the input variables sequentially with their minimum and maximum uncertainty.

As a second step a Monte Carlo error estimation is performed for MODIS TIT to account for the combination of the uncertainties of the input variables and their impact on the retrieved TIT. 2401 MODIS scenes of winter 2007/08 and 2240 MODIS scenes of winter 2008/09 are used.

The TIT retrieval is a non-linear function of the input variables:

$$h_i = h_i(T_s, T_a, U_{10m}, RH) \tag{23}$$

Table 9: Overview of the uncertainties in the input variables that are used for the calculation of the thin-ice thickness. The uncertainties are based on the mentioned references. © Table: Adams et al. [2012], Table 4.

	MODIS	NCEP
Ice-surface temperature (°C)	±1.6 (Hall et al. [2004])	
2-m air temperature (°C)		±4.5 (Ernsdorf et al. [2011], Renfrew et al. [2002])
10-m wind speed (m/s)		±1.3 (Ernsdorf et al. [2011], Renfrew et al. [2002])
Relative humidity (%)		±20 (Renfrew et al. [2002])

For each MODIS scene 100 random values within the error range of each variable are generated (Table 9). Then, a population of 100 synthetic TIT distributions is calculated using MODIS and NCEP input variables plus a random error value. Since the focus is on the ice production only combinations with a negative net energy flux to the atmosphere Q_A and unstable stratification are used (see Section 4.2). Due to these restrictions the number of TIT calculations can be less than 100 for some MODIS pixels (89 on average).

4.4.1.2 Results and discussion

The first part of the statistical analysis gives insight into the ranking of the input variables in terms of their impact on the retrieved ice thickness. The analysis focuses on TIT up to 20 cm for clarity. Moreover, the results of the Monte Carlo error estimation show that a maximum limit of 20 cm ice thickness is a better choice than 50 cm. The errors in the ice thickness become very large when the ice is thicker than 20 cm. Table 10 shows the reference polynya area (including all TIT pixels up to 20 cm) and the anomaly in percent when varying the input variables about their minimum and maximum uncertainty. The averaged reference polynya area (POLA) is 8,780 km². The highest anomalies are found for T_a: an under-/overestimation of T_a results in +154/-57 % pixels up to 20 cm of ice thickness, respectively.

The results with varying uncertainties indicate that T_a has the highest (change in POLA of +141/-55 %) and RH the lowest impact (change in POLA of +10/-8 %) on the derived TIT and hence polynya area (POLA). T_s has the second highest impact on the calculated ice-thickness (change in POLA of +75/-44 %). The hierarchy is the following: (1) T_a, (2) T_s, (3) U_{10m} and (4) RH.

Table 10: Ranking of the input variables in terms of their impact on the retrieved polynya area (thin ice ≤ 20 cm). The averaged polynya area (POLA) of 10 MODIS scenes (Table 7) and the anomaly in percent when varying the input variables (T_s = MODIS ice-surface temperature, T_a = 2-m air temperature, RH = relative humidity, U_{10m} = 10-m wind speed) about their minimum and maximum uncertainty is shown (Table 9). The thin-ice thickness is calculated with MODIS T_s and NCEP atmospheric variables. © Table: Adams et al. [2012], Table 5.

Variable	TIT POLA (km²)	T_s +1.6 °C	T_s -1.6 °C	T_a +4.5 °C	T_a -4.5 °C
average / change	8,780	+75%	-44%	-55%	+141%
Variable	RH +20%	RH -20%	U_{10m} +1.3 m s⁻¹	U_{10m} -1.3 m s⁻¹	
average / change	-8%	+10%	+22%	-23%	

After the examination of POLA, the impact of the input variables' uncertainties with regard to different ice classes is investigated. Figure 16a shows the average number of pixels of the reference ice thickness for the three ice classes 0-5 cm (cl1), 5-10 cm (cl2) and 10-20 cm (cl3). The other subplots of Figure 16b-i present the absolute anomaly to the reference ice thickness varying each input variable by about its minimum and maximum uncertainty. The uncertainties in the retrieved TIT up to 5 cm are smallest for each variable (e.g., +66/-42 % for T_s). Higher anomalies are found for the ice class 10-20 cm for all input variables indicating an increase in TIT uncertainty with thicker ice (e.g., +72/-44 % for T_s). The highest anomalies are found for T_a: an underestimation of T_a results in +154 % more and an overestimation of T_a in -57 % less thin-ice pixels between 10 and 20 cm of thickness. This is in agreement with the analysis of POLA where the overestimation of POLA is highest when T_a is underestimated. The error analysis of the ice classes confirms that T_a and T_s uncertainties have the highest impact; RH and U_{10m} uncertainties have only a minimal impact on the retrieved TIT.

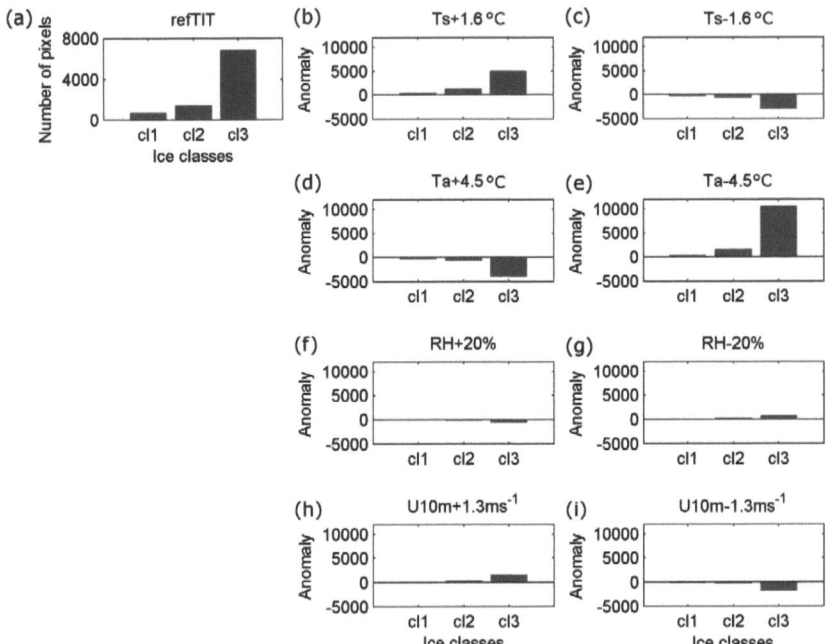

Figure 16: (a) Averaged sum of reference thin-ice pixels for ice classes cl1 0-5 cm, cl2 5-10 cm and cl3 10-20 cm. (b)-(i) Anomaly to reference ice thickness for the input variables with their minimum and maximum uncertainty (Table 9). © Figure: Adams et al. [2012], Figure 4.

The results of the Monte Carlo error estimation for different ice classes and the winter seasons 2007/08 and 2008/09 are shown in Table 11. Comparing the uncertainty values for the two years, it is found that up to 20 cm of ice thickness the errors are low and almost the same (errors of ±1 to ±5 cm). For thicker ice the uncertainties become larger (errors of ±12 to ±60 cm) and differ strongly between the two winter seasons (24 cm difference for the ice class 40-50 cm). A further explanation about this feature is given in Section 4.4.2.

The mean uncertainty in the TIT from 0 to 20 cm is ±4.7 cm for the winter 2007/08 and ±4.6 cm for winter the 2008/09; the mean over both seasons is ±4.7 cm. For TIT from 0 to 50 cm the mean uncertainty is ±26.1 cm for the winter 2007/08 and ±36.0 cm for winter the 2008/09; the mean over both seasons is ±31.1 cm.

The results of the Monte Carlo error estimation lead to the conclusion that for subsequent application only MODIS TIT up to 20 cm should be used. In terms of ice production, TIT in this range are sufficient since it is assumed that ice formation rates are strongest in regions with ice thickness below 20 cm (Ebner et al. [2011]).

Table 11: Statistical uncertainty calculated with Monte Carlo error estimation using the uncertainties listed in Table 9 for winter 2007/08 and 2008/09 and different ice classes. © Table: Adams et al. [2012], Table 6.

Ice class (cm)	Winter 2007/08 $TIT_{MODIS+NCEP}$ (cm)	Winter 2008/09 $TIT_{MODIS+NCEP}$ (cm)	Mean of both winters $TIT_{MODIS+NCEP}$ (cm)
0 - 5	±1.0	±1.0	±1.0
5 - 10	±2.0	±2.2	±2.1
10 - 20	±5.2	±5.3	±5.3
20 - 30	±16.8	±12.0	±14.4
30 - 40	±34.2	±28.4	±31.3
40 - 50	±36.7	±60.2	±48.5
mean 0 - 20	±4.7	±4.6	±4.7
mean 0 - 50	±26.1	±36.0	±31.1

In conclusion, the results of the statistical sensitivity analysis show:

(1) The mean absolute error for both winter seasons is ±4.7 cm for ice of 0-20 cm of thickness and ±31.1 cm for ice of 0-50 cm of thickness;

(2) the uncertainties in T_s and T_a strongly influence the calculation of the thin-ice thickness;

(3) an underestimation of T_a results in a strong underestimation of the thin-ice thickness (i.e. more pixels with TIT below 20 cm and hence larger polynya areas); an overestimation of T_a results in an overestimation of the thin-ice thickness but the anomaly to the reference value is moderate;

(4) the uncertainties in the atmospheric variables have a smaller impact on very thin ice (up to 10 cm) than on thicker ice;

(5) for subsequent applications only MODIS thin-ice thickness below 20 cm should be used.

4.4.2 Comparison of ice-thickness data sets using different atmospheric data

The statistical analysis provides the mean error of the calculated thin-ice thickness. The comparative sensitivity analysis investigates the uncertainties which can emerge from different atmospheric data sets. Here, the atmospheric data sets NCEP and COSMO data are used. The reasons for using COSMO in addition to NCEP as an atmospheric data set are:

(1) Ernsdorf et al. [2011] show that COSMO data better agree with observations at the fast ice compared to NCEP;

(2) the high resolution COSMO simulations include the impact of the polynyas on the atmospheric boundary layer.

Two selected case studies are analyzed and discussed in the following. The focus is on the TIT up to 20 cm (see Section 4.4.1). Additionally, Envisat ASAR images are used for comparison.

The first case study was chosen for the 3 January, 2009 0135 UTC (Figure 17). At this date, the Anabar-Lena polynya is partly open. The polynya opens on 30 December, 2008 and closes on 7 January, 2009. MODIS T_s resolves the polynya structure with a sharp temperature transition between fast ice and polynya area. Adjacent to the fast-ice edge a small band of relatively high temperatures (around -8 °C) occur indicating an area of very thin ice. Going further off-shore the surface temperature decreases continuously. The NCEP T_a field does not show a polynya signal (no heating of the lower atmosphere). Air temperatures are almost homogenously distributed (T_a around -22 °C) throughout the whole Laptev Sea. In contrast, the COSMO T_a field shows higher temperatures (up to -15 °C) in polynya regions and an abrupt change to very cold temperatures (around -22 °C) going further off-shore. However, the COMSO T_a pattern is not the same as for MODIS T_s.

Contour lines for TIT of 10 and 20 cm are indicated in Figure 17a,c. The total polynya area with thin ice less than 20 cm is approximately the same for $TIT_{MODIS+NCEP}$ (8054 km^2) and $TIT_{MODIS+COSMO}$ (6675 km^2) and consistent with the polynya area shown in Envisat ASAR backscatter coefficients (Figure 17e).

In the following, the main problems resulting from the use of coarse resolution NCEP data in combination with MODIS T_s are examined. In addition, the case study also shows how the high-resolution COSMO data affects on MODIS TIT. When interpreting $TIT_{MODIS+COSMO}$, the differences between polynya area from the remote sensing data sets MODIS and AMSR-E must be taken into account.

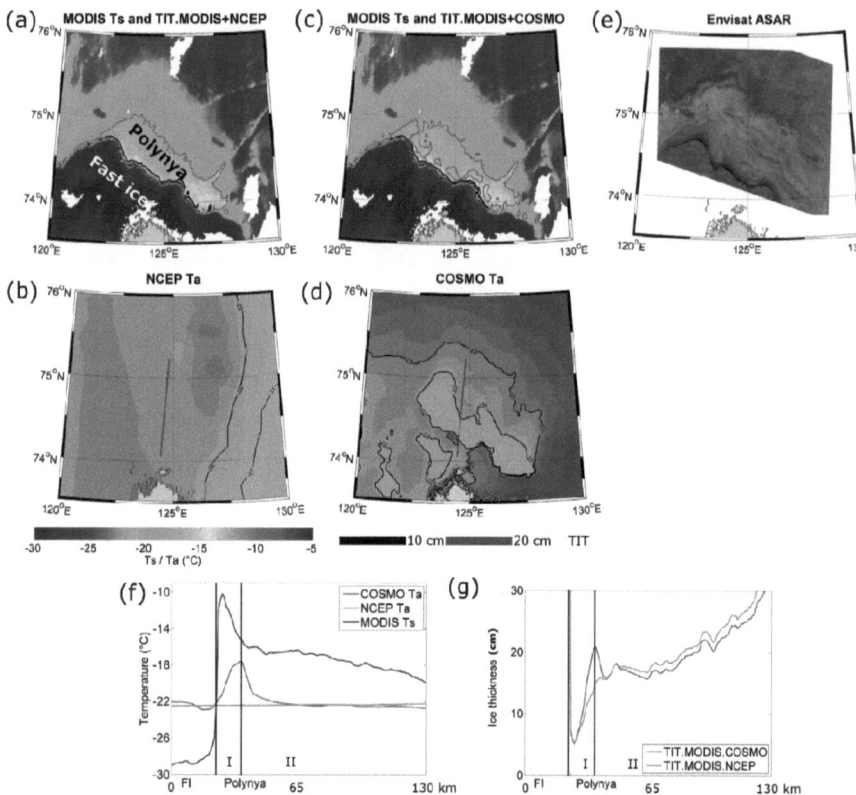

Figure 17: (a) MODIS ice-surface temperature (T_s) from 3 January 2009 0135 UTC with ice-thickness (TIT) contour lines of $TIT_{MODIS+NCEP}$ product; (b) corresponding NCEP 2-m air temperature (T_a) from 3 January 2009 0000 UTC; (c) MODIS T_s from the same date with ice thickness contour lines of $TIT_{MODIS+COSMO}$ product; (d) corresponding COSMO T_a from 3 January 2009 0200 UTC; Red line in (b) and (d) marks the transect shown in (f) and (g); (e) Envisat ASAR image of 3 January 2009 0243 UTC (f) NCEP COMSO T_a and MODIS T_s for the transect across the polynya; (g) ice thickness as given by $TIT_{MODIS+NCEP}$ and $TIT_{MODIS+COSMO}$ for the transect across the polynya. FI = fast ice. © Figure: Adams et al. [2012], Figure 5.

The interdependence of T_s and T_a and the impact on the retrieved TIT is illustrated with the across-polynya transects shown in Figure 17f,g. Comparing the temperatures and the corresponding TIT the transects can be divided into two domains. In domain I, consistent temperature peaks of COSMO T_a and MODIS T_s are observed starting at the transition region between fast ice and very thin ice. NCEP T_a is constant over the entire transect and up to 4 °C lower than COSMO T_a. In spite of the difference in NCEP and COSMO T_a the resulting ice thickness starts with very thin ice for both retrievals. This can be explained by the strong temperature gradient between the surface and the lower atmosphere (see Equation 18) in regions of very thin ice leading to similar TIT mostly independent of the atmospheric data sets. Due to these strong temperature differences the uncertainties of the atmospheric input variables are masked. Since the quality of MODIS T_s is high (accuracy of ±1.6 °C) and taking this masking effect into account, TIT are reliable up to 10 cm more or less independent of the uncertainties in the atmospheric variables for this case study. This is supported by the statistical sensitivity analysis (Figure 16; Table 11). With increasing distance from the fast-ice edge (end of domain I) the peak in COSMO T_a leads to thicker ice in $TIT_{MODIS+COSMO}$ in comparison to $TIT_{MODIS+NCEP}$. COSMO T_a is more reliable because the modification of the atmospheric boundary layer due to polynyas, which results in higher air temperatures, is taken into account. NCEP T_a is underestimated resulting in underestimated TIT. This means that atmospheric data sets which do not capture polynyas generally underestimate T_a in these regions as well as underestimate the calculated thin-ice thickness.

In domain II, MODIS T_s indicate a gradual thickening of the ice. However, $TIT_{MODIS+COSMO}$ shows thinner ice. The ice thickness for both TIT products is almost identical. This indicates that COSMO and NCEP T_a are both too low. While the inconsistency between MODIS T_s and NCEP T_a is obvious (too coarse spatial resolution, no polynyas), the explanation for the COSMO data is given by inconsistencies between the AMSR-E based polynya mask used in COSMO and the MODIS data. The polynya signal in COSMO simulations results from AMSR-E SIC used as a lower boundary and the model's fine spatial resolution (5 km) which is able to resolve polynyas. In the COSMO model, all grid cells with AMSR-E SIC below 70 % are set to 10 cm ice thickness; all other sea-ice grid cells are set to 1 m ice thickness. This jump from thin ice to thick ice results in the relatively abrupt temperature decrease of COSMO T_a downstream of the temperature peak (Figure 17d,f).

In the following the differences between polynya area derived from MODIS and AMSR-E are investigated in more detail. Although ice thickness is derived from AMSR-E (Martin [2005]) these methods rely on a calibration using MODIS TIT (see Section 2.2.5). Thus, SIC from AMSR-E is used. In accordance with other studies it is assumed that all grid cells with AMSR-E SIC below 70 % describe the polynya area (e.g., Massom et al. [1998], Adams et al. [2011]; see Section 2.2.4). In order to investigate the ability of AMSR-E to capture polynyas, Figure 18 compares AMSR-E polynya area (POLA) and $TIT_{MODIS+NCEP}$ POLA for a complete winter period. The results indicate that in general the AMSR-E POLA capture polynyas of an ice thickness between 10 and 15 cm. Polynya regions including higher ice thickness are mostly missing, particularly for mid-winter months. In the view of these results, it is reasonable that a mean ice thickness of 10 cm is used for polynya areas in COSMO. However, in regions of thicker ice not classified as a polynya COSMO T_a gives lower temperatures than expected in these regions. A transition zone of continuously decreasing temperatures, as seen in MODIS T_s, is therefore not represented in COSMO T_a, resulting in the temperature jump mentioned above.

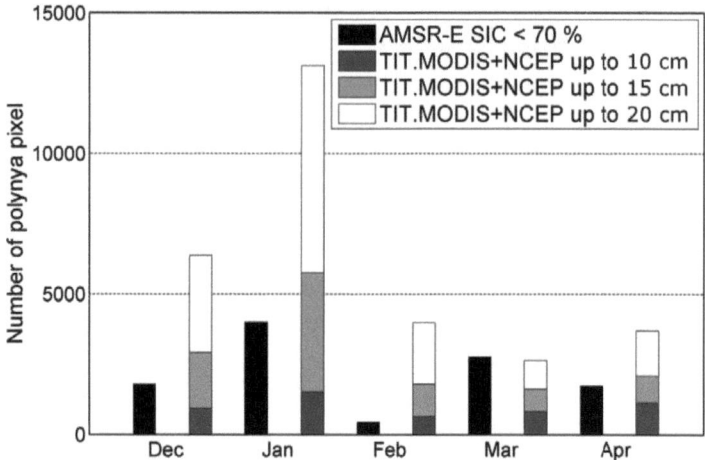

Figure 18: Number of polynya pixels counted from daily $TIT_{MODIS+NCEP}$ composites up to 10 cm (dark grey), 15 cm (light grey) and 20 cm (white) as well as the number of polynya pixels as counted from AMSR-E sea-ice concentration (SIC) lower than 70 % (black) for December, 2007 to April, 2008. MODIS and AMSR-E are interpolated to the 5 km COSMO grid. The MODIS daily composites cover on average approximately 70 % of the Laptev Sea polynya (see Section 4.6). Only pixels that are covered by both data sets are used for the calculation. © Figure: Adams et al. [2012], Figure 6.

The differences in the MODIS T_s and the AMSR-E data result from the different measurement techniques of both data sets (see Section 2.1). At 1 km × 1 km, MODIS data has a finer spatial resolution than the AMSR-E data (6.25 km × 6.25 km). The coarse spatial resolution of AMSR-E data has an impact on the quality of the ice-concentration retrieval. Pixels between fast/drift ice and open water/thin ice are influenced by both ice types resulting in mixed pixels as stated by Willmes et al. [2010]. In particular, this is important for narrow polynyas. Due to these characteristics, AMSR-E POLA is generally underestimated in comparison to $TIT_{MODIS+NCEP}$ POLA. An adjustment of the SIC threshold from 70 % to a higher value does not yield better results, but leads to false classification in non-polynya regions. It is concluded that COSMO data includes flaw polynyas, whose sizes are underestimated.

The second case study shows a polynya in the Anabar-Lena region on 6 January, 2009 0205 UTC (Figure 19). MODIS T_s is around -5 °C adjacent to the fast-ice edge and decreases going further off-shore. Again, NCEP T_a is homogenously distributed (around -15 °C). COSMO gives T_a around -17 °C in the polynya region and around -23 °C in the fast-ice and drift-ice regions and shows again the abrupt temperature decrease at the off-shore polynya border. As was also observed in the first case study, the polynya seen in COSMO T_a is here narrower than the polynya signal in MODIS T_s.

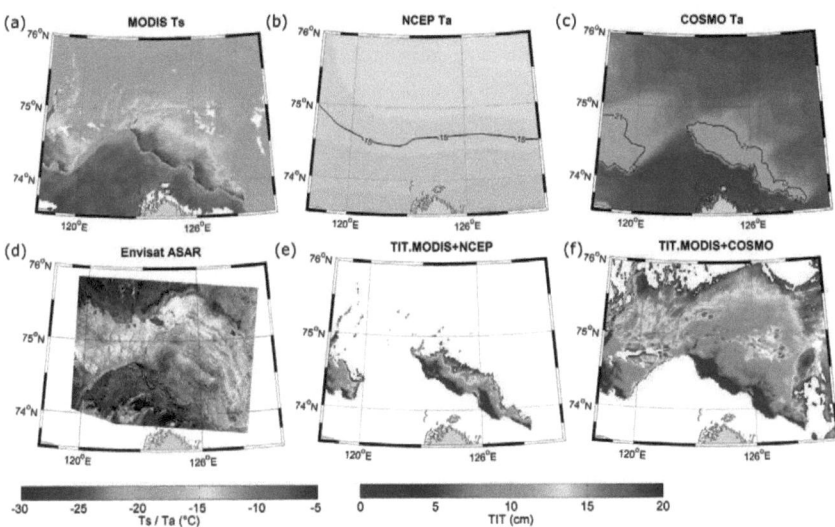

Figure 19: (a) MODIS ice-surface temperatures (T_s) from 6 January 2009 0205 UTC; (b) corresponding NCEP 2-m air temperature (T_a) from 6 January 2009 0000 UTC; (c) corresponding COSMO T_a from 6 January 2009 0200 UTC; (d) Envisat ASAR image of 6 January 2009 0248 UTC; (e) ice-thickness distribution as calculated with MODIS T_s and NCEP atmospheric variables ($TIT_{MODIS+NCEP}$); (f) ice-thickness distribution as calculated with MODIS T_s and COSMO atmospheric variables ($TIT_{MODIS+COSMO}$). © Figure: Adams et al. [2012], Figure 7.

Although no polynya is captured in NCEP, T_a is warmer in all regions (fast ice, polynya, drift ice) than COSMO T_a. In the polynya region, NCEP T_a is around 2 °C warmer; in fast-ice and drift-ice regions NCEP T_a is around 8 °C higher than COSMO T_a. The NCEP warm bias occurs in relation to COSMO for the two winter seasons 2007/08 and 2008/09 with a mean bias of 1 °C and 3 °C, respectively. If NCEP T_a is overestimated, thin-ice thickness retrievals with MODIS T_s and NCEP atmospheric variables will result in overestimated TIT

(Figure 19e). In contrast, the calculated thin ice-thickness with MODIS T_s and COSMO atmospheric variables are thinner for this case study (Figure 19f). The area with TIT less than 20 cm is extremely large for $TIT_{MODIS+COSMO}$ (large polynya: 38,456 km^2) in comparison to $TIT_{MODIS+NCEP}$ (small polynya: 7,862 km^2). The contrary impact on the derived ice thickness using overestimated or underestimated T_a is significant. In agreement with the statistical analysis (Figure 16; Table 10) the case study shows that an underestimation of T_a has a strong influence on the retrieved ice thickness (TIT are extremely underestimated).

Taking this warm bias of NCEP T_a into account, the difference of the errors for the two winters for ice thicker than 20 cm can be explained (Monte Carlo error estimation, see Section 4.4.1). The difference results from errors in the NCEP data. The number of dates with underestimated or overestimated NCEP T_a during polynya events must be different in the two years resulting in different TIT errors. However, a masking effect of the uncertainty in the atmospheric variables is discovered for very thin ice due to the strong temperature gradient between surface and lower atmosphere (see Section 4.4.2), the differences in the uncertainty between the two winter seasons are therefore firstly obvious for thicker ice.

The Envisat ASAR image for the same date shows a polynya signature. However, a conclusion cannot be drawn about the TIT distribution from Envisat ASAR to decide which TIT distribution is more reliable.

Summarizing the results from the two case studies, four situations concerning uncertainties of the atmospheric variables and their effect on the retrieved ice thickness can be identified:

(1) COSMO has a polynya temperature signal which corresponds to the MODIS T_s polynya signal (see case study of Willmes et al. [2010]): highest accuracy of the resulting TIT retrieval;

(2) COSMO has a polynya signal, but the polynya is narrower than the one in MODIS data: high accuracy in the area where COSMO variables are adjusted to the polynya, excessively thin ice in the area where MODIS T_s still shows a polynya signal and COSMO T_a is underestimated (inconsistency at the T_a jump from thin ice to thick ice);

(3) COSMO (if the polynya is extremely narrow and not resolved in AMSR-E SIC) and NCEP do not show a polynya signal (T_a underestimated): the retrieved ice thickness is underestimated;

(4) NCEP T_a is overestimated (warm bias): the retrieved ice thickness is overestimated.

4.4.3 Nearly-coincident MODIS T_s and TIT as an uncertainty indicator

After the statistical and comparative sensitivity analysis the uncertainty in TIT is determined using a third method. First, pairs of MODIS scenes that meet the following three criteria are identified for the two winter seasons 2007/08 and 2008/09: (1) open polynya, (2) only a 20-minute time differences and (3) partial spatial overlap. This allows the data to account for the uncertainty in the TIT distributions by examination of the difference of the two distributions with such a short time lag. Using the difference as an uncertainty indicator is appropriate because the thermodynamic ice-thickness change over 20 minutes should be negligible. Schröder et al. [2011] analyzed the ice growth rate in the second half of April, 2008 for a model study. According to the study the maximum growth rate per day is 27 cm. This equates to an average growth rate of 0.38 cm per 20 minutes.

The previous sensitivity analyses focused: (1) on the average TIT error including all input variables and (2) on the impact of the 2-m air temperature error on the TIT uncertainty. In comparison, this analysis reveals information about the uncertainty in the TIT resulting from MODIS T_s. The NCEP data used in combination with MODIS T_s for the calculation of the ice thickness is available at 6 h intervals, implying that the TIT distributions with the 20-minute time lag are retrieved using the same NCEP data but different MODIS T_s. The difference between the two TIT distributions must occur due to altering MODIS T_s.

Figure 20 shows a case study for the 5 January, 2009. The first MODIS scene was detected at 0415 UTC, the second one at 0435 UTC. MODIS T_s (T_s1 and T_s2) and the TIT distributions (TIT1 and TIT2) for both times are presented as well as the T_s difference and TIT difference. The differences are calculated subtracting the first data set from the second one.

At the date of the case study, the Anabar Lena polynya is wide open with considerable areas of open water and very thin ice. T_s1 and T_s2 as well as TIT1 and TIT2 are similar to each other. They differ slightly near the drift-ice edge. Upon closer inspection of the T_s difference, a small band along the fast-ice edge with large differences is found. The differences are mostly positive, around +15°C, but in some parts negative, around -8°C. In all other regions the T_s differences are between -2 and 0 °C.

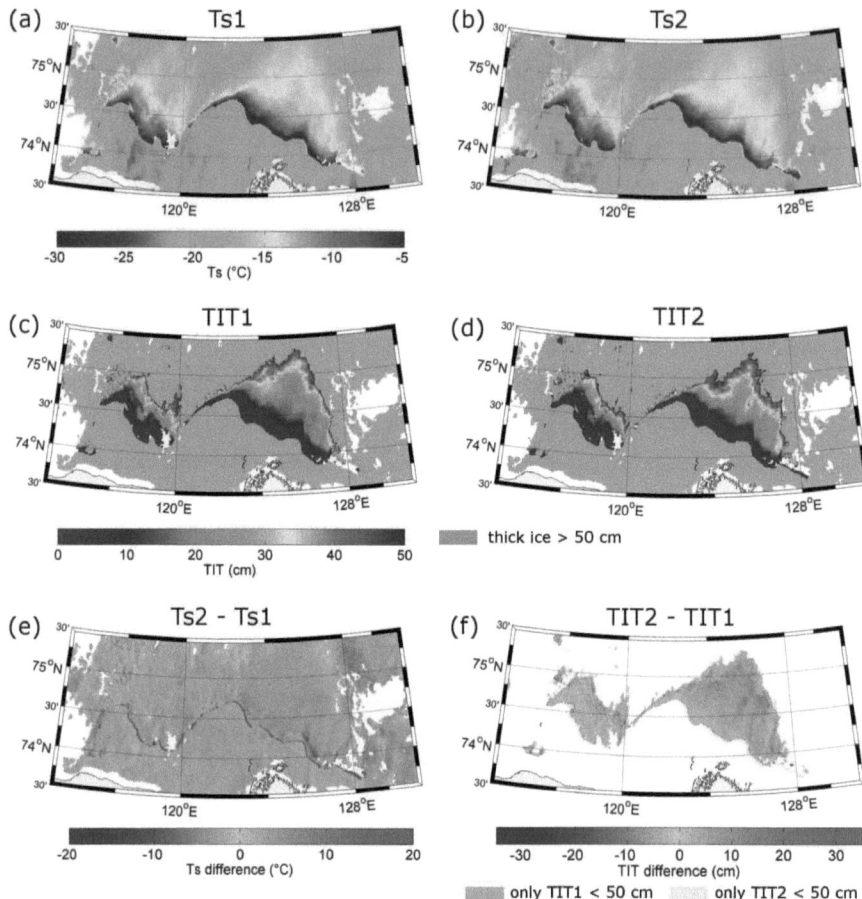

Figure 20: Example for deriving the uncertainty in the thin-ice thickness using nearly-coincident MODIS images. (a) MODIS ice-surface temperatures (T_s1) from 5 January, 2009 0415 UTC. (b) MODIS T_s2 from 5 January, 2009 0435 UTC. (c) Thin-ice thickness distribution (TIT1) from the first date and time. (d) TIT2 from the second date and time. (e) Difference between T_s1 and T_s2 (positive numbers denote temperature increase, negative numbers denote temperature decrease). (f) Difference between TIT1 and TIT2 (positive numbers denote ice growth, negative numbers denote decreased ice thickness). The blue-gray-red color map shows only the pixels that are covered by TIT1 $\leq$ 50 cm and TIT2 $\leq$ 50 cm. The green color marks the area where TIT1 shows ice thickness $\leq$ 50 cm and TIT2 shows ice thickness > 50 cm. For the yellow color it is vice versa.

The TIT difference (TIT2-TIT1) shows a small band of TIT pixels along the fast-ice edge that only occur in the TIT2 distribution (coincident with the areas of strong positive T_s differences). In these regions, warm T_s2 results in TIT2. In the regions of strong negative T_s differences (T_s1 is relatively warmer than T_s2), only TIT1 is retrieved.

Table 12: An overview of the 6 examples used for the determination of the uncertainty in the thin-ice thickness by nearly-coincident MODIS images.

Date	First time	Second time	Average wind speed (m s^{-1})	Average wind direction
26 December 2007	0500	0520	6.59	SSE
27 December 2007	0230	0250	4.18	SSE
28 December 2007	0310	0330	3.93	SO
30 December 2007	0300	0320	8.63	SSE
29 December 2008	0230	0250	7.93	SW
5 January 2009	0415	0435	9.78	SW

The TIT difference is approximately zero between the fast-ice edge, and regions up to approximately 12 cm. In regions of thicker ice the TIT differences increase up to +20 cm although the T_s difference remains constant. This indicates that the relation between T_s and TIT is strong for very thin ice and decreases with increasing ice thickness.

The T_s difference map shows mostly values between 0 and -2 °C, which is in agreement with the uncertainty in T_s of ±1.6 °C (Hall et al. [2004]). The higher differences along the fast-ice edge might be the result of the different satellite measurement geometry of the two MODIS scenes and small errors in the geolocation of the pixels. The geolocation accuracy of the MODIS data is determined to within ±150 m and with an operational goal of ±50 m at nadir (Nishihama et al. [1997]). Due to the satellite measurement geometry (ascending or descending) it is possible that not exactly the same areas are covered by the fields of views of the two MODIS scenes. Thus, the information used for the average determination of T_s in a field of view could be different for the two MODIS scenes. This means that in regions of large temperature differences, as they occur at the fast-ice edge, the average determination of T_s for a pixel can be more influenced either by the warm polynya temperatures or the cold fast-ice temperatures resulting in the small shift of the fast-ice edge between the two scenes.

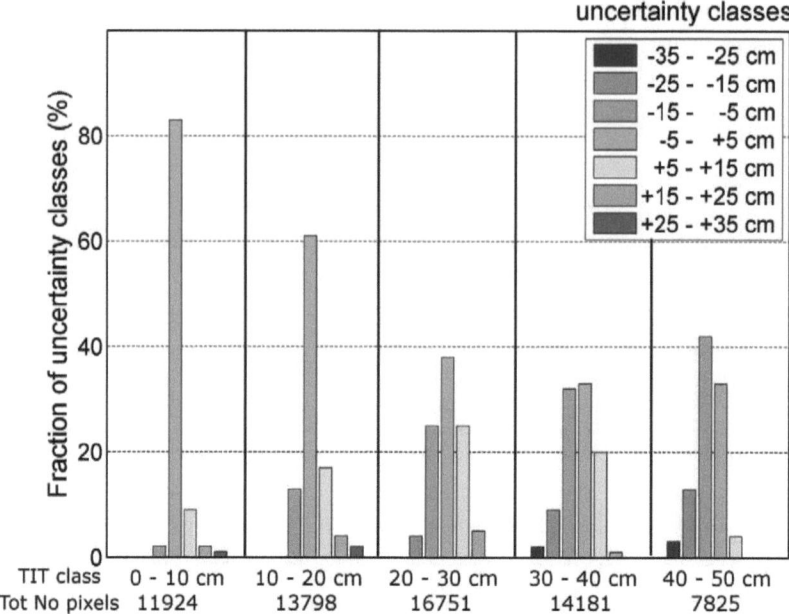

Figure 21: Polynya regions which are covered by the thin-ice distributions of images are selected for the 6 examples listed in Table 12. The thin-ice area is divided into five thin-ice thickness (TIT) classes from 0 to 50 cm (10 cm bins). At the lower part of the figure the total number of pixels counted for each ice class is denoted (based on TIT from the first time, sum of 6 examples). The difference (= uncertainty) between the two TIT distributions ranges from -35 to +35 cm. Divided into 10 cm bins the frequency distribution of the uncertainty is given in percent for each ice class.

4.4.4 Summary and conclusion of the sensitivity analyses

The sensitivity analysis study is divided into three parts. In the first part, the average uncertainty in the thin-ice thickness is determined based on a statistical method. The second part refers to the uncertainties in T_a since this variable has the greatest impact amongst the atmospheric variables on the resulting TIT. The effect of uncertainties in MODIS T_s on the resulting TIT is highlighted in the third part.

The statistical error estimation is performed with MODIS T_s and NCEP atmospheric variables for the two winter seasons 2007/08 and 2008/09. Firstly, the ranking of all input variables is analyzed. It is found that the uncertainties of T_a and T_s have the highest impact on the derived TIT distribution. Secondly, a Monte Carlo error estimation is performed for all MODIS scenes in the two winter seasons. The errors between both winters differ for ice thicker than 20 cm, indicating the high influence of the uncertainty in NCEP T_a. The mean absolute error for ice thickness up to 20 cm is ±4.7 cm.

The comparative sensitivity analysis shows two case studies with NCEP and COSMO T_a as well as the derived thin-ice products $TIT_{MODIS+NCEP}$ and $TIT_{MODIS+COSMO}$. For the higher quality atmospheric data set, COSMO, the results show that the prescription of the sea-ice field using AMSR-E SIC gives polynyas that are too narrow when compared to MODIS T_s. Due to the inconsistencies between MODIS and AMSR-E polynya areas, an underestimation of the ice thickness occurs with increasing distance from the fast-ice edge. It is concluded that a simple prescription of the polynya area as done in the COSMO model configuration is not sufficient for a realistic simulation of the atmospheric variables. A prescription of the TIT distribution in atmospheric models as provided by the $TIT_{MODIS+NCEP}$ data set is required to describe the polynya situation more accurately.

In contrast, NCEP atmospheric data does not capture polynyas, but shows a homogenously distributed T_a. If the NCEP T_a is underestimated the retrieved TIT is underestimated. However, the warm bias in NCEP T_a found in comparison to COSMO T_a leads to rather overestimated TIT.

The effect of the uncertainties in MODIS T_s on TIT is analyzed with coincident overpasses of different satellites. The results are good for TIT below 20 cm. However, along zones of strong temperature differences, such as the fast-ice edge, the error in MODIS T_s (and hence in the TIT distribution) could be larger due to the different satellite measurement geometry of the two scenes and errors in geolocation.

As an overall conclusion of the sensitivity analyses, it can be demonstrated that:

(1) the uncertainty in very thin ice is mostly independent of the uncertainties in the atmospheric variables and MODIS T_s;

(2) an underestimation of T_a results in a strong underestimation of the ice thickness and an overestimation of T_a results in a moderate overestimation of the ice thickness;

(3) the error in MODIS T_s affects strongly the TIT in zones of strong temperature differences and in areas of thicker ice. In the latter regions, the uncertainty is larger than ±1.6 °C.

Due to these reasons, the NCEP data set is appropriate for the calculation of TIT up to 20 cm in the Laptev Sea for the two winter seasons 2007/08 and 2008/09. Moreover, the combination of MODIS T_s with NCEP atmospheric variables is applicable for the calculation of TIT up to 20 cm for longer time periods (2000-today) and other polar regions. When applying the energy balance model to longer periods, problems with the MODIS cloud mask could appear. During the night, MODIS visible channels cannot be used to identify clouds, reducing the quality of the mask (Ackerman et al. [1998], Frey et al. [2008], Ackerman et al. [2008]). The TIT studies are therefore only focused on sea-ice regions where polynyas occur (application of a polynya mask; see Section 4.2). This helps to reduce TIT artifacts.

$TIT_{MODIS+NCEP}$ for the two winters represent a valuable high-resolution verification data set for sea-ice simulations and remote sensing methods. However, a complete spatial coverage of sea-ice areas is rare because of the limitations by cloud coverage and nighttime conditions.

4.5 Transfer of the thin-ice thickness algorithm to other regions

In the previous section, the thermal-infrared thin-ice thickness retrieval is applied for the Laptev Sea and a sensitivity analysis is performed for this region. To test the application of the improved algorithm, the retrieval is transferred to another Arctic region, the Lincoln Sea, and an Antarctic region, the Weddell Sea.

4.5.1 Lincoln Sea (Arctic)

The Lincoln Sea is an Arctic shelf sea located between the north-east coast of Ellesmere Island (Canada) and the north-west coast of Greenland (Figure 22). In the south, the Lincoln Sea borders the Nares Strait.

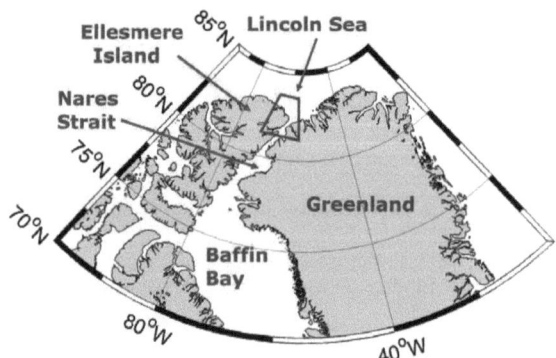

Figure 22: Overview map of the Lincoln Sea and adjacent regions. Red box denotes the Lincoln Sea.

The polynya evolution in the Lincoln Sea as hypothesized by Kozo [1991] is schematically illustrated in Figure 23. In autumn, an ice arch forms in the funnel-shaped Lincoln Sea and blocks the ice from the north. During northerly wind conditions, the ice breaks up at the southern border of the ice arch and is advected southward through the Nares Strait. This polynya formation process is typical for a latent heat polynya (see Section 1.1). Barber and Massom [2007] categorize the Lincoln Sea polynya as an ice-bridge polynya. If no ice bridge is formed, the sea ice will be advected from the Lincoln Sea through the Nares Strait. The marginal sea-ice zone will not turn into a polynya region with new ice formation until an ice arch forms.

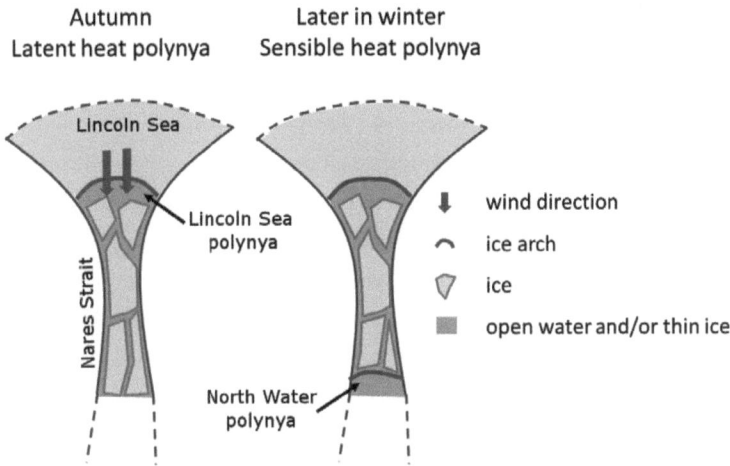

Figure 23: Formation processes of the Lincoln Sea polynya.

Kozo [1991] further suggests that later in winter, a second ice arch forms at the southern end of the Nares Strait implying a change of polynya evolution in the Lincoln Sea. The ice flow through the Nares Strait stops and the Lincoln Sea polynya should freeze over. However, historical data documents that the polynya is open during large parts of the winter season (Kozo [1991]). The reason for the polynya opening might be the tidally induced vertical mixing of heat from deeper water to the surface. This polynya type is called sensible heat polynya (see Section 1.1). Due to the two different evolutionary processes within one winter season, the Lincoln Sea polynya is named hybrid polynya (Kozo [1991]).

In previous studies, the ice conditions in the Lincoln Sea are investigated in combination with the sea-ice outflow through the Nares Strait and the ice-arch dynamics (Kwok [2005], Kwok et al. [2010]). Moreover, Haas et al. [2006] analyzed ice-thickness profiles detected by a helicopter-borne electromagnetic measurement system in the Lincoln Sea. The result of their study is that multi-year ice dominates the thickness distribution but first-year ice is also found, representing the ice formed in the polynya.

The study of Kozo [1991] is the first one explicitly dealing with the Lincoln Sea polynya. Extensive monitoring studies using remote sensing data as performed for many Arctic and Antarctic polynya systems are not available. Gudmandsen [2000] is the only scientist to have carried out studies examining the occurrence of polynyas in the Lincoln Sea using AVHRR, SSM/I and European Remote Sensing Satellite (ERS) Synthetic Aperture Radar (SAR) data. In a subsequent study, Gudmandsen [2005] investigated the Lincoln Sea polynya and ice drift with RADARSAT imagery from the winter seasons 1996/97 and 1997/98. Based on the two studies, they tried to prove the hypothesis of a hybrid polynya. They investigated that the polynya occurs every winter from September/October until April, but they are indecisive concerning the evolutionary process of the polynya. Additional atmospheric and oceanographic data is necessary to investigate properly the polynya formation process. This indicates research demand for the Lincoln Sea polynya concerning its evolution, the spatial and temporal variability of polynya size and hence the new ice formation.

As a first step towards answering these questions, the thermal-infrared thin-ice thickness retrieval is applied to the Lincoln Sea. The two presented case studies show MODIS T_s, NCEP T_a, NCEP wind speed and direction, the resulting TIT distribution as well as an Envisat ASAR and an AMSR-E SIC image for comparison. The first case study of the 1 May, 2008 2220 UTC is illustrated in Figure 24. At this time, a wide ice arch is formed between Ellesmere Island and Greenland. The Envisat ASAR image shows a sharp border where the ice arch occurs. This structure is not seen in MODIS T_s.

NCEP T_a shows the same shortcomings in the Lincoln Sea as in the Laptev Sea. The temperature is homogenously distributed and the temperature above the polynya is underestimated.

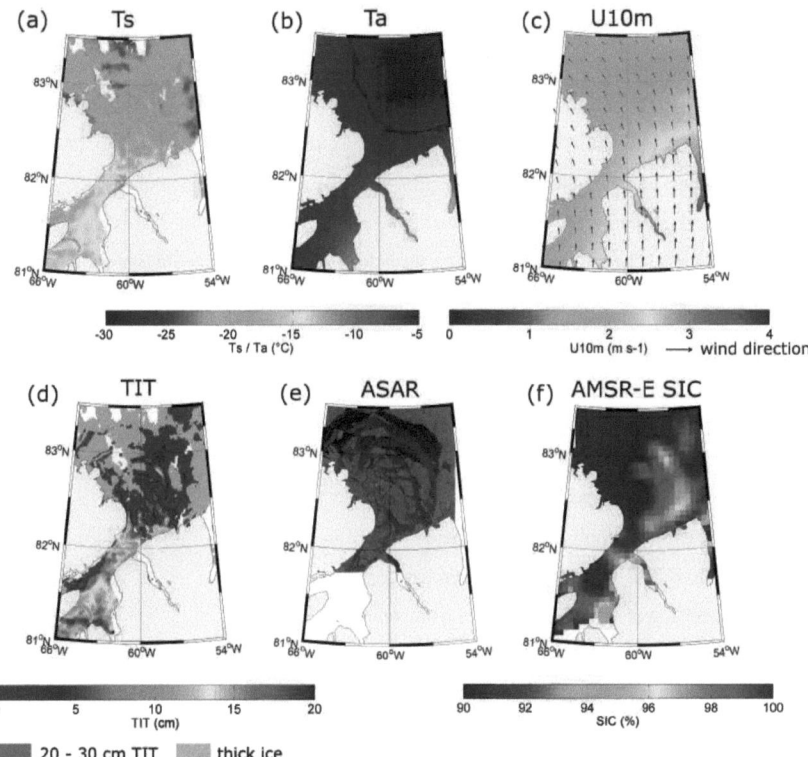

Figure 24: Case study of the Lincoln Sea polynya of 1 May, 2008 2220 UTC. (a) MODIS ice-surface temperature (T_s) detected at this date and time. (b) NCEP 2-m air temperature (T_a) of 2 May, 2008 0000 UTC. (c) NCEP 10-m wind speed (U_{10m}) (colormap) and wind direction (black arrows) of 2 May, 2008 0000 UTC. (d) Thin-ice thickness (TIT) retrieved with MODIS T_s and NCEP atmospheric variables on 1 May, 2008 2220 UTC. (e) Envisat ASAR image detected of 1 May 2008 1709 UTC. (f) Daily AMSR-E sea-ice concentration (SIC) of 1 May, 2008. Note that the colormap shows SIC values from 90 to 100 %. White areas in (a) and (d) result from the MODIS cloud mask.

The TIT distribution shows some thin ice below 20 cm of thickness in the Lincoln Sea and all over the northern part of the Nares Strait (polynya size = 8434 km^2). The small structures shown by Envisat ASAR are mostly visible in the MODIS data. In contrast, the AMSR-E SIC shows no polynya but SIC reduced to 96 %. Following the hypothesis of Kozo [1991], this polynya should be a sensible heat polynya. The southerly wind at this date could support that theory.

The second case from 19 April, 2009 0000 UTC is illustrated in Figure 25. In 2009, a small ice arch formed near the Nares Strait. The Envisat ASAR image clearly shows the blocked ice north of the ice arch and the ice break-up south of it. The two different ice zones are found in the MODIS T_s image, as well. The surface temperature of the thick ice blocked by the ice arch is below -20 °C. The polynya in the Lincoln Sea is represented by relatively warm temperatures. Again, the NCEP T_a is homogenously distributed and the temperature in the polynya region is underestimated. The thin-ice distribution shows mostly ice thickness between 0 and 20 cm in the Lincoln Sea and the northern part of the Nares Strait (polynya size = 4667 km^2). The AMSR-E SIC is partly below 70 % in the Lincoln Sea (polynya size = 1431 km^2).

In contrast to the first case study, a large, relatively homogenous area covered by thin ice occurs, allowing the coarser resolution AMSR-E data to capture this area with reduced SIC. Consistently low wind speeds from the south-west support the hypothesis that the opening of the polynya is not wind-driven.

The conclusions from these two case studies are:

(1) it is feasible to apply the TIT retrieval to the Lincoln Sea;
(2) AMSR-E SIC is only reliable when larger polynyas occur;
(3) the position of the ice arch varies from year to year.

To get a better overview of the polynya dynamics in the Lincoln Sea polynya, the MODIS TIT retrieval should be applied for the last 12 years. Due to the gaps in the MODIS TIT distribution, the fully-coverage daily AMSR-E SIC can be used to derive the polynya area for the last 10 years in the Lincoln Sea. However, one has to note that small-scale structures as shown on 1 May, 2008 are not resolved by AMSR-E.

In order to clarify the polynya evolution, the ice-arch formation in the Lincoln Sea and Nares Strait has to be further investigated. Barber and Massom [2007] state that the processes responsible for the formation of the ice arches are not well understood. Figure 26 shows the position of the ice arches in the Lincoln Sea and Nares Strait for 13 years and indicates that the position and number of the arches vary from year to year. Due to the ice-arch dynamics, the Lincoln Sea, the Nares Strait and the North Water should be monitored as one system. Furthermore, oceanographic data is necessary in order to investigate the exchange processes in the different ocean layers (tidal mixing). Wind speed, wind direction and ocean currents should be observed for longer periods of time.

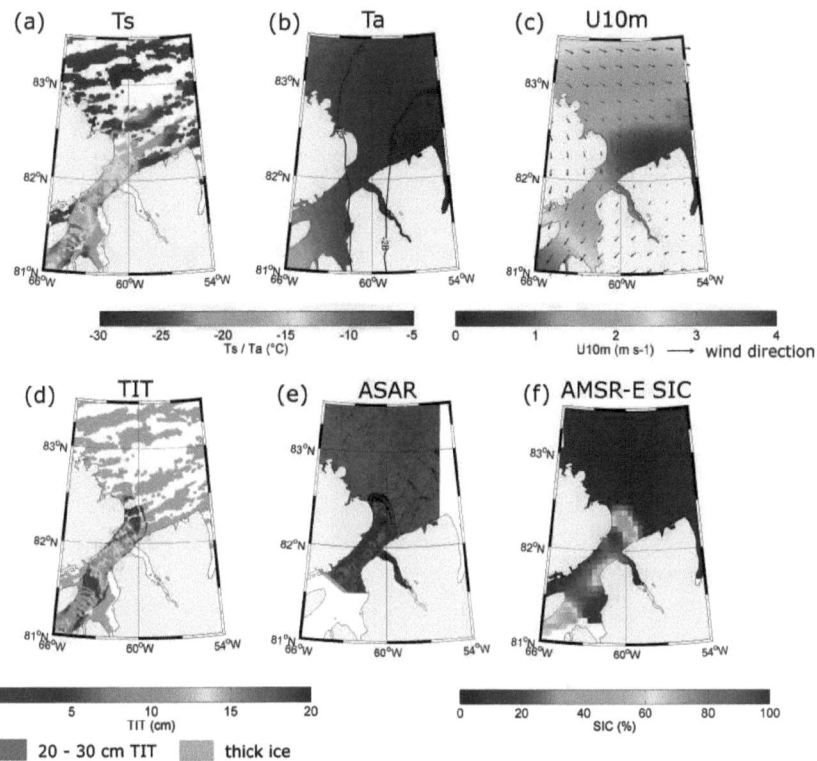

Figure 25: Case study of the Lincoln Sea polynya of 19 April, 2009 0000 UTC. (a) MODIS ice-surface temperature (T_s) detected at this date and time. (b) NCEP 2-m air temperature (T_a) at the same date and time. (c) NCEP 10-m wind speed (U_{10m}) (colormap) and wind direction (black arrows) at the same date and time. (d) Thin-ice thickness (TIT) retrieved with MODIS T_s and NCEP atmospheric variables. (e) Envisat ASAR image is detected on 18 April, 2009 2245 UTC. (f) Daily AMSR-E sea-ice concentration (SIC) on 19 April, 2008. Note that the colormap shows SIC values from 0 to 100 %. White areas in (a) and (d) result from the MODIS cloud mask.

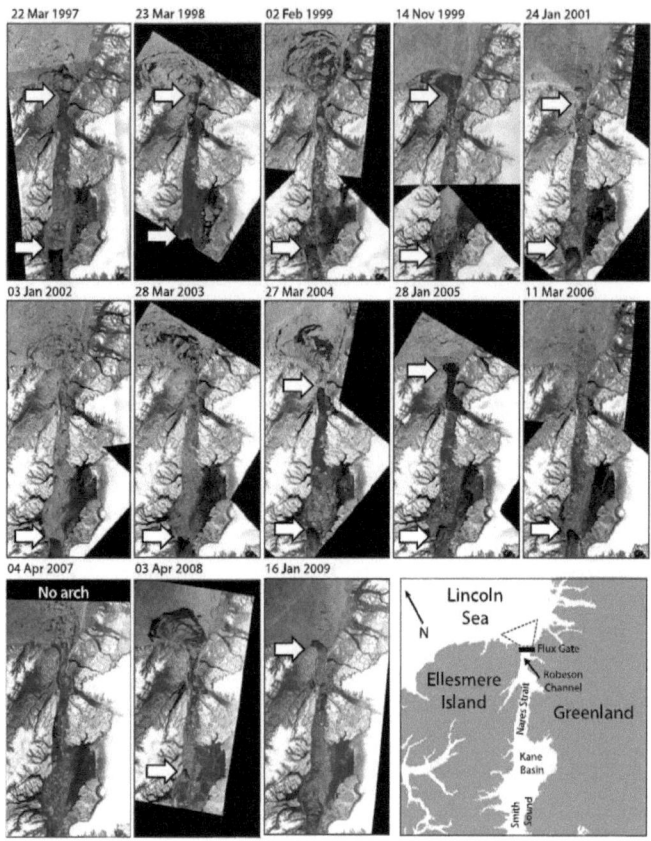

Figure 26: RADARSAT and Envisat ASAR images showing the ice situation in the Lincoln Sea and Nares Strait for 1997-2009. The white arrows denote the position of the ice arches. © Figure: Kwok et al. [2010], Figure 1.

4.5.2 Weddell Sea (Antarctic)

The Weddell Sea is a marginal sea of the Southern Ocean (Figure 27). The western border of the Weddell Sea is the Antarctic Peninsula and the southern border the Ronne-Filchner Ice Shelf. In the north, the Weddell Sea is limited by the Atlantic-Indian ocean ridge and in the east by Coats Land. The Weddell Sea has the largest winter sea-ice cover of the Antarctic shelf seas with 4.75×10^6 km^2 (Drucker et al. [2011]). Coastal latent heat polynyas driven by low pressure systems occur along the Ronne-Filchner Ice Shelf and the east coast (Comiso and Gordon [1998]). According to Markus et al. [1998], the Ronne-Filchner polynya is the region in the Weddell Sea where the largest ice production occurs. The Weddell Sea polynyas contribute significantly to the production of Antarctic Bottom Water (AABW) (Marsland [2004], Van Woert [1999]).

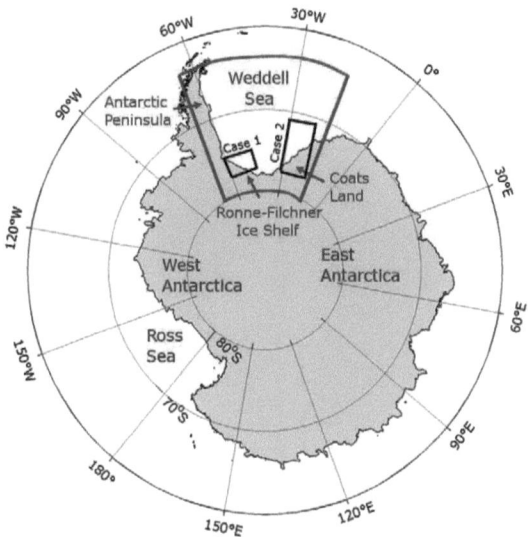

Figure 27: Overview map of the Antarctic. Red box denotes the Weddell Sea. Black boxes mark the position of the polynyas presented in the two case studies shown in Figure 28 and Figure 29.

In the past, a few remote sensing studies were performed to investigate polynya dynamics and ice production in the Antarctic shelf seas. Tamura et al. [2008] used SSM/I data (with help of thermal-infrared TIT) to derive the polynya area (see Section 2.2.5 for description of the method) and hence the ice production in the Antarctic coastal polynyas. For the Weddell Sea, they estimated an ice production of 84.6 ±25 km^3 with a trend of -30 km^3 (10yr)$^{-1}$. Kern [2009] used the polynya signature simulation method (PSSM) to retrieve polynya area in the marginal seas of the Antarctic continent (see Section 2.2.6 for description of the method). He calculated an average polynya area for 1992-2008 of 212,000 km^2 for the central Weddell Sea and 220,000 km^2 for the eastern Weddell Sea.

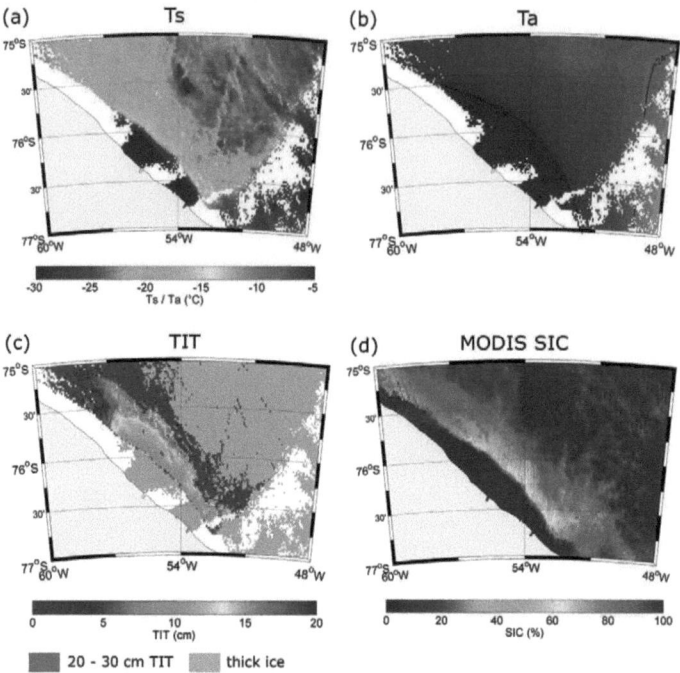

Figure 28: Case study 1 of the Weddell Sea polynya on 14 April, 2008 0305 UTC. (a) MODIS ice-surface temperature (T_s) detected at this date and time. (b) NCEP 2-m air temperature (T_a) at the same date at 0600 UTC. (c) Thin-ice thickness (TIT) retrieved with MODIS T_s and NCEP atmospheric variables. (d) Daily MODIS sea-ice concentration (SIC) of 14 April, 2008.

Friedrich [2010] studied the long-term polynya dynamics in the Weddell Sea from 1979 to 2009. She used SSM/I and AMSR-E SIC with a 70 % threshold (Massom et al. [1998]) to derive polynya area. Additionally, TIT retrievals following Tamura et al. [2007], Martin [2004] and Martin [2005] were applied (see Section 2.2 for description of the methods). The TIT results calculated by Tamura et al. [2007] agree well with the results of Kern [2009].

Moreover, Friedrich [2010] addressed the spatial resolution issue when SSM/I and AMSR-E polynya area is compared. AMSR-E polynya area is two times smaller than SSM/I polynya area. Due to the higher spatial resolution of AMSR-E data, the polynya detection is more accurate and the influence of mixed pixels that are counted as polynya area is lower. Based on this, Friedrich [2010] suggests that the polynya area and thin-ice thickness distribution based on SSM/I are questionable. Therefore, polynya studies based on high-resolution satellite data are essential for the investigation of the Weddell Sea.

Extensive polynya studies using MODIS TIT have not been performed for the Weddell Sea until now. In this study, the improved TIT retrieval is applied to the Weddell Sea for case studies. The first case study on 14 April, 2008 0302 UTC, illustrated in Figure 28, shows a polynya along the Ronne-Filchner Shelf Ice. The comparison between MODIS T_s, SIC and TIT indicates that the TIT retrieval works well for the Weddell Sea. The polynya covers a large area but is completely covered with a thin-ice layer.

Regarding the NCEP T_a, one can see that the air temperature is not modified due to the polynya. This drawback of the NCEP data is described in detail in Section 4.4.2. The fact that NCEP data does not capture polynyas results in underestimated T_a above the polynya indicating underestimated TIT.

The second case study on 20 September, 2008 0205 UTC is presented in Figure 29. At this time, a long narrow polynya was present along the east coast of the Weddell Sea. Again, the TIT distribution indicates that the polynya is completely covered by thin ice and no open water is detected. As mentioned for the first case study, TIT is probably underestimated due to the underestimated NCEP T_a above the polynya. MODIS TIT and SIC show similar polynya structures.

The two case studies indicate the feasibility of the TIT retrieval application to the Weddell Sea. As mentioned above, cloud identification by the MODIS cloud algorithm, especially during the night, is often not precise in regions of thin clouds and sea smoke (Ackerman et al. [1998], Frey et al. [2008]). The cloud coverage in the Antarctic is very high. For instance, the east Antarctic cloud cover fraction in October 2003 was 93 % (Spinhirne [2005]). In comparison, the global average cloud fraction was 70 % (Spinhirne et al. [2004]). This is a critical factor when applying the TIT algorithm to the Weddell Sea polynya. Thin-ice regions that can be visually classified as cloud structures appear in many scenes. To reduce the problem, the application of a polynya mask as for the Laptev Sea is necessary. A second method is to create thin-ice thickness composites to reduce the cloud influence (see Section 4.6 and Fraser et al. [2009]). Moreover, working towards an improved MODIS cloud mask is necessary to solve the cloud problem.

It can be concluded that the thin-ice thickness algorithm is applicable to the Weddell Sea, but one has to be careful with misclassification of polynya area due to clouds that are not identified by the MODIS cloud mask (sea smoke, thin clouds, etc.).

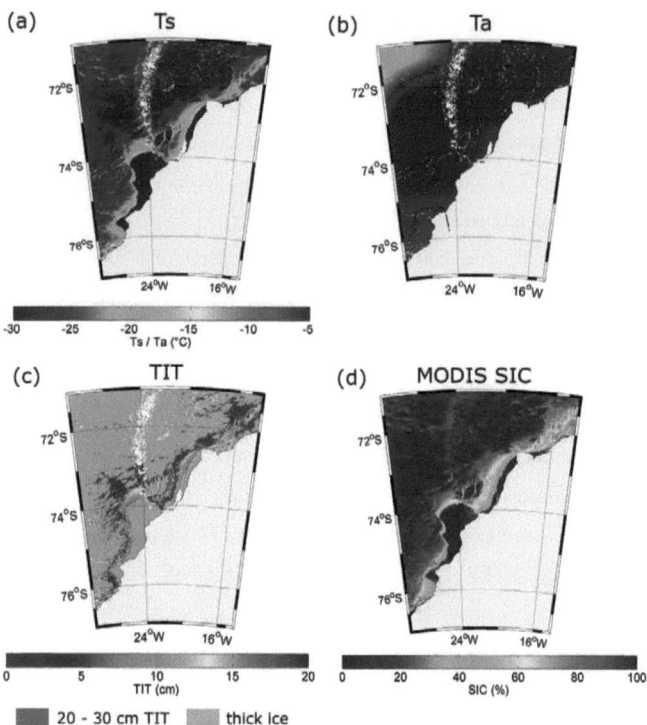

Figure 29: Case study 2 of the Weddell Sea polynya on 20 September, 2008 0205 UTC. (a) MODIS ice-surface temperature (T_s) detected at this date and time. (b) NCEP 2-m air temperature (T_a) at the same date at 0000 UTC. (c) Thin-ice thickness (TIT) retrieved with MODIS T_s and NCEP atmospheric variables. (d) Daily MODIS sea-ice concentration (SIC) of 20 September, 2008.

4.6 Retrieval of further MODIS products

4.6.1 Daily thin-ice thickness maps

The knowledge presented by the sensitivity analyses in Section 4.4 is the basis of the computation of daily thin-ice thickness (TIT) maps. The following results revealed are important when the TIT composite is generated:

- the $TIT_{MODIS+NCEP}$ product can be used for long time studies;
- the highest accuracy with an average uncertainty of ±4.7 cm is determined for ice thickness between 0 and 20 cm;
- the error of the ice class 20-30 cm is ±14.4 cm;

- a polynya mask is applied to the single TIT scenes excluding the central Laptev Sea and hence TIT patches occurring due to clouds.

For the computation of daily TIT maps, $TIT_{MODIS+NCEP}$ up to an ice thickness of 30 cm are used. For this method, the single MODIS scenes are provided without the polynya mask. If necessary, the polynya mask is applied to the final daily TIT composite.

Mostly, MODIS single scenes do not cover the Laptev Sea completely. This is due to the restriction that only cloud-free T_s pixels can be used for the TIT retrieval. Moreover, due to the satellite measurement geometry the single MODIS images cover often only parts of the Laptev Sea polynya. In order to achieve a better coverage, daily TIT composites are computed. Additionally, correction schemes are implemented to reduce the impact of incorrect TIT patches (e.g., clouds, underestimated NCEP T_a).

These issues motivate a calculation of a daily MODIS TIT data set. The calculation consists of four steps:

(1) Correction of the single MODIS TIT images using MODIS T_s

As mentioned above, the usage of NCEP atmospheric data can result in inaccurate ice thickness. The comparison of thin-ice distributions with Envisat ASAR images indicates that an underestimation of TIT due to an underestimated NCEP T_a can be observed in some cases.

To reduce the impact of incorrect TIT patches, a MODIS T_s difference threshold is applied. According to the calculation of MODIS SIC following Drüe and Heinemann [2004], a background surface temperature within a 200 × 200 pixel kernel is generated. After that, the difference of each MODIS T_s pixel in this kernel is determined. If the difference temperature is lower than the predefined difference threshold of 2.5 °C, the ice thickness is set to 30 cm. In the following, the corrected single thin-ice thickness distribution is referred to as TIT_{corrTs}. The original thin-ice distribution as calculated by the TIT retrieval is abbreviated as TIT_{org} in the following.

(2) Calculation of a daily thin-ice composite

A daily thin-ice thickness map consists of all available MODIS scenes per day, 15 on average. Due to missing values and the variability of the polynya size during a day, the number of valid pixels which are used per day for the determination of the composite varies between 1 and 18 (highest numbers during polar night).

Before starting the conservative composite determination, all ice thicknesses above 30 cm are excluded. For each pixel, the median is determined from the available thin-ice thicknesses per day. The median-value compositing is preferable to maximum/minimum/mean-value compositing. As a result of the median-value, TIT patches due to unidentified clouds are reduced.

(3) Correction of daily TIT composites using MODIS SIC

In very few cases, the conservative composite determination causes TIT areas to disappear in the daily composite. To correct this, daily MODIS SIC maps are used (Drüe and Heinemann [2004], Drüe and Heinemann [2005]). If a pixel in a daily composite is not classified as thin ice below 30 cm but has a SIC less than 70 %, the pixel is set to 30 cm ice thickness.

(4) Application of a polynya mask due to the cloud problem

The last step includes a manual verification of all daily TIT composites to correct for the presence of clouds not identified by the MODIS cloud mask. Although the MODIS T_s product includes a quality-controlled cloud mask, sea smoke or thin clouds are often not detected by the cloud-mask algorithm resulting in overestimated T_s (Ackerman et al. [1998], Hall et al. [2004]). In these cases, the MODIS T_s represents the cloud temperature rather than the surface temperature. If this temperature is similar to the polynya T_s, the clouds are misinterpreted as thin-ice regions. Due to this, some T_s images show polynya regions in the central Laptev Sea that can definitively be identified as cloud structures. Because the efforts to improve the cloud identification are insufficient, the daily composites are proofed manually and a polynya mask is applied to exclude the central part of the polynya if necessary (see Figure 1; Bareiss and Görgen [2005]). This minimizes the error due to clouds in regions where no polynyas occur.

For winter 2007/08, the polynya mask is applied to 40 out of 152 composites, for winter 2008/09 to 36 from 151 composites.

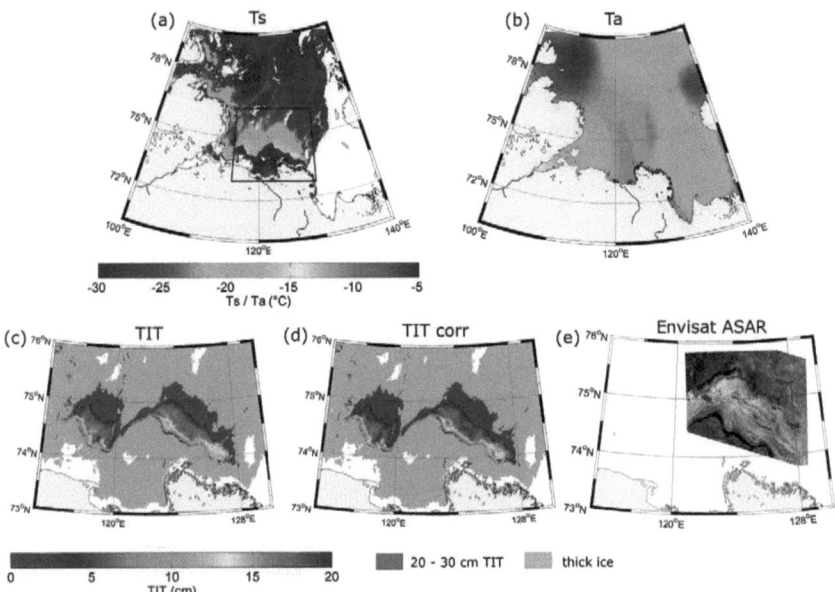

Figure 30: Case study of 3 January, 2009 0135 UTC. (a) MODIS ice-surface temperature (T_s). Black bordered area denotes subsets shown in (c)-(e). (b) NCEP T_a. (c) MODIS thin-ice thickness (TIT) as a result of the thin-ice thickness retrieval used with MODIS T_s and NCEP data. (d) MODIS TIT after the correction by means of the MODIS T_s difference threshold. (e) Envisat ASAR image of 3 January, 2009 0243 UTC.

The final thin-ice thickness data set consists of three classes: (1) ice thicknesses between 0 and 20 cm; (2) a separate bulk class ice for ice thicknesses between 20 and 30 cm; (3) ice thicker than 30 cm. On average, the thin-ice thickness maps cover 59 % of the entire Laptev Sea and 67 % of the Laptev Sea polynya.

Up to now, this method is applied for the MODIS scenes detected in the months from December to April of the two winter seasons 2007/08 and 2008/09. As MODIS data is available since 2000, the product can be provided for the last 12 years.

Following the description of the different steps to determine the daily MODIS thin-ice thickness composite, two case studies are presented. The first case study for 3 January, 2009 0135 UTC shows an open Anabar-Lena polynya (Figure 30). The polynya size of TIT_{org} is 14,878 km^2 and of TIT_{corrTs} 13,464 km^2. Because TIT_{org} and TIT_{corrTs} images do almost not differ from each other, it is concluded that the temperature difference threshold correction has almost no effect on this case study (correction step 1). This can be explained by means of the maps of MODIS T_s and NCEP T_a. These two variables mainly impact the resulting thin-ice thickness. The NCEP T_a has a warm bias at this date in the Laptev Sea region. This avoids the calculation of underestimated ice thickness and hence overestimation of the polynya area. The visual comparison with the Envisat ASAR image detected on the 3 January, 2009 0243 UTC shows that the polynya size and edges are reliable.

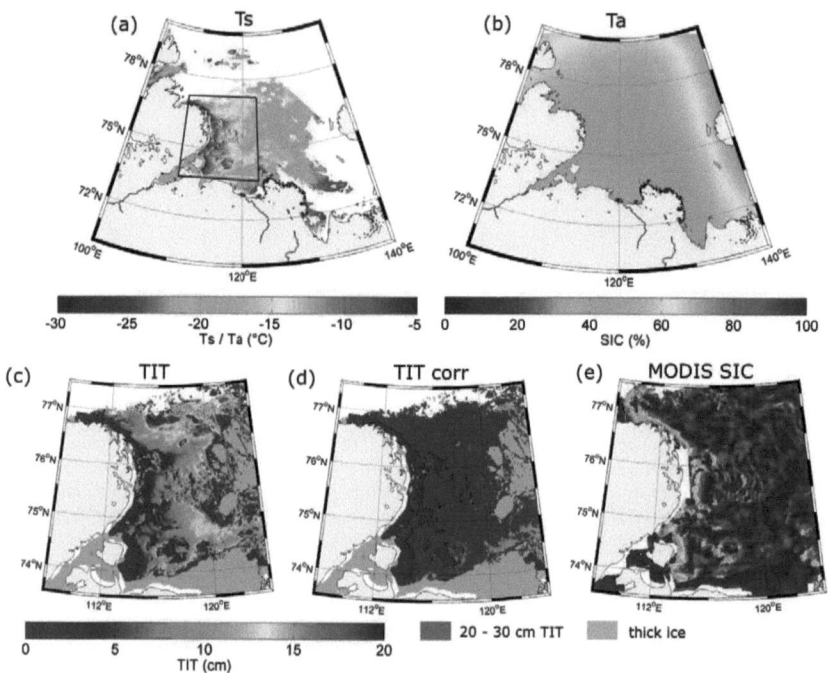

Figure 31: Case study of 13 January, 2009 2025 UTC. (a) MODIS ice-surface temperature (T_s) Black bordered areas denotes subsets shown in (c)-(e). (b) NCEP T_a. (c) MODIS thin-ice thickness (TIT) as a result of the thin-ice thickness retrieval used with MODIS T_s and NCEP data. (d) MODIS TIT after the correction by means of the MODIS T_s difference threshold. (e) Daily MODIS sea-ice concentration (SIC) of 13 January, 2009. All SIC < 70 % are polynya area.

The second case study presents MODIS T_s, NCEP T_a as well as TIT_{org} and TIT_{corrTs} from the 13 January, 2009 2025 UTC (Figure 31). At this date, the Taimyr polynya is open. The comparison of the two TIT maps shows large differences. The polynya size of TIT_{org} is 85,090 km^2 and of TIT_{corrTs} 22,902 km^2. After the correction, the polynya (TIT_{corrTs}) is smaller in comparison to the TIT_{org}. The examination of T_s and T_a implies that the combination of the relatively warm T_s and the colder T_a yields TIT less than 20 cm. The comparison with

MODIS SIC shows that TIT_{corrTs} fits the polynya edges given by MODIS SIC (polynya means SIC < 70 %).

Figure 32 shows the resulting daily MODIS TIT composite for 3 January, 2009 and 13 January, 2009. Regarding the daily composite on 13 January, a thin-ice area in the central Laptev Sea occurs. After applying the polynya mask, the data is excluded in the central Laptev Sea. On the one hand, this reduces the coverage of the data set but on the other hand the quality of the data set is increased. Although for some cases the application of a polynya mask reduces the coverage of the product, the average coverage of the entire Laptev Sea is still 59 %.

The retrieved daily TIT product is a valuable data set for verification of other model and remote sensing ice-thickness data. The quantification of the retrieval error for different ice classes allows the assimilation of the data into sea-ice/ocean models.

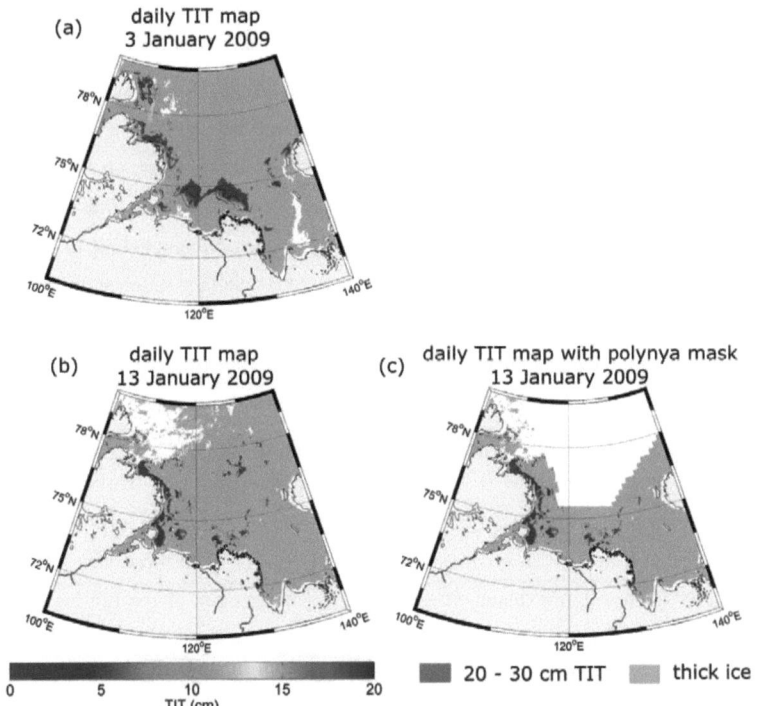

Figure 32: Daily MODIS thin-ice thickness (TIT) composites for (a) 3 January, 2009 and (b) 13 January, 2009 without polynya mask. TIT patches appear in the center of the Laptev Sea due to unidentified clouds. (c) Daily MODIS TIT composites for the 13 January, 2009 with polynya mask.

4.6.2 Monthly fast-ice masks

The fast-ice variability in the Laptev Sea needs to be investigated in more detail. The understanding of many processes and linkages to other environmental factors is poor and requires further examination. Motivated by this, monthly fast-ice masks are derived from high-resolution MODIS imagery.

Before introducing the method to derive fast-ice masks, the fast-ice situation in general and in the Laptev Sea is described and the existing remote sensing based methods are shortly reviewed.

Fast ice is consolidated sea ice that is attached to a shore and does not move with ocean currents or winds (World Meteorological Organization [1990], Mahoney et al. [2007]). The formation of fast-ice can be thermodynamic (Heil [1999]) or dynamic (Massom et al. [2001], Massom et al. [2009]). Thermodynamic fast-ice growth is generally limited to small shallow bays and narrow straits (Antonova [2011]). Dynamic processes include the attachment of ice floes to the coast or existing fast ice and fast-ice break-ups due to ocean currents and wind.

Persistent and annually recurring fast ice is an important element of the coastal regions in the Arctic's marginal seas. The fast-ice layer as an interface between ocean and atmosphere inhibits the exchange processes between them and affects the two layers on a regional scale (Massom et al. [2001]). The position of flaw polynyas is controlled by the fast-ice extent and shape (Massom et al. [2001]). Fast ice also influences the coastal morphology (e.g., reducing the wave-based coastal erosion, Lantuit and Pollard [2008]) and the sediment transport (e.g., potential sediment entrainment area, Eicken et al. [2000]). According to Bareiss and Görgen [2005], fast ice is affected by the fresh-water cycle (e.g., the fresh-water inflow could have an impact on the position of the fast-ice edge in the subsequent winter). Furthermore, fast ice is biologically important as a habitat for microorganisms and large animals (Krell et al. [2003]). Finally, the fast-ice distribution is crucial for marine navigation and off-shore exploration (Johannessen [2005], Hughes et al. [2011]).

In the Laptev Sea, fast ice shows seasonal variability. It begins to form along the coast in October and reaches its maximum extent in April. The position of the fast-ice edge then coincides roughly with the position of the 25 m isobaths (see Figure 1, Figure 34). Fast-ice decay starts at the end of May (Bareiss [2003], Bareiss and Görgen [2005]). At its maximum extent, the fast ice reaches approximately 400-500 km off the coast (Bareiss and Görgen [2005], Dmitrenko et al. [1999]). The largest extent of fast ice is found in the eastern Laptev

Sea with approximately 150,000 km^2. In the western Laptev Sea, the fast ice covers approximately 60,000 km^2 (Project 'System Laptev Sea' [2011]).

The forcing processes for fast-ice formation are not completely understood. Therefore, Antonova [2011] investigated the fast-ice forcing parameters in the south-eastern Laptev Sea for the years 2003-2011. Firstly, she found a strong linkage between the local bathymetry and the position, shape, and extent of the fast ice (Lieser [2004]). Secondly, off-shore wind conditions are important for fast-ice formation. When the fast ice is fully developed, small-scale variability can be wind-driven. A third connection was found between fast ice and the large-scale atmospheric circulation. A linkage between fast ice and (anti)cyclonic activity in January, February and April as well as between fast ice and atmospheric vorticity in some years was revealed.

Obtaining in-situ measurements in the fast-ice regions is extremely difficult and not practicable during polar night (Project 'System Laptev Sea' [2011]). Thus, monitoring and investigation of the fast ice with remotely sensed data is crucial. Additionally, this data provides good spatial and temporal coverage to examine the fast-ice variability.

Efforts in fast-ice retrieval are undertaken using microwave (passive/active) and optical (visible/infrared) satellite imagery. Passive microwave satellite data is convenient for fast-ice monitoring due to its insensitivity to clouds but it lacks sufficient spatial resolution. Selyuzhenok [2011] used AMSR-E T$_B$ at 89 GHz as a first approach and applied a correlation technique similar to Fraser et al. [2010]. Fast-ice areas are identified by high correlation coefficients found in a time series of consecutive images. A drawback is the misclassification of fast ice within drift-ice regions. The usage of additional data sets could help to make the fast-ice extent more reliable. For instance, a bathymetric limitation would exclude the spurious fast ice in the drift-ice area (Selyuzhenok [2011]).

SAR data has a high spatial resolution and is not influenced by clouds but its temporal and spatial coverage is insufficient. Mahoney et al. [2004] developed an automated method using SAR data and applied it to the Alaskan Arctic. However, fast ice often has the same normalized backscatter as drift ice and wind roughened open water (Lythe et al. [1999]). More information can probably be retrieved when multi-channel, double-polarized SAR data is used.

Fraser et al. [2009] generates 20-day composites of MODIS visible and infrared imagery to remove the cloud cover in the east Antarctic. After the composite computation, Fraser et al. [2010] use a correlation technique to identify fast-ice regions (see above). In spite of the compositing of MODIS imagery, the imperfect cloud mask forces the use of manual classification of the cloud-free MODIS images. Additionally, AMSR-E data is used to close gaps resulting from persistent cloud formations.

In summary, no recent method using remote sensing data can automatically provide a fast-ice data set with full coverage and fine spatial and temporal resolution.

In this section, two methods are presented to derive monthly fast-ice masks from MODIS data. As a first approach, MODIS T_s is used to determine monthly fast-ice masks from December, 2007 to April, 2008 (Adams et al. [2011]). A suitable MODIS scene (cloud-free above the fast-ice edge) is sought for each of the five months. The temperature differences between the thicker and colder fast ice and the thinner and warmer drifting ice are utilized (see Figure 10). A dynamic ice surface temperature threshold separating fast ice from drifting ice is defined for each selected MODIS scene. Manual post-processing is necessary. Figure 33 shows the temporal and spatial variability of the fast ice in the Laptev Sea from December, 2007 to April, 2008 as derived from MODIS T_s.

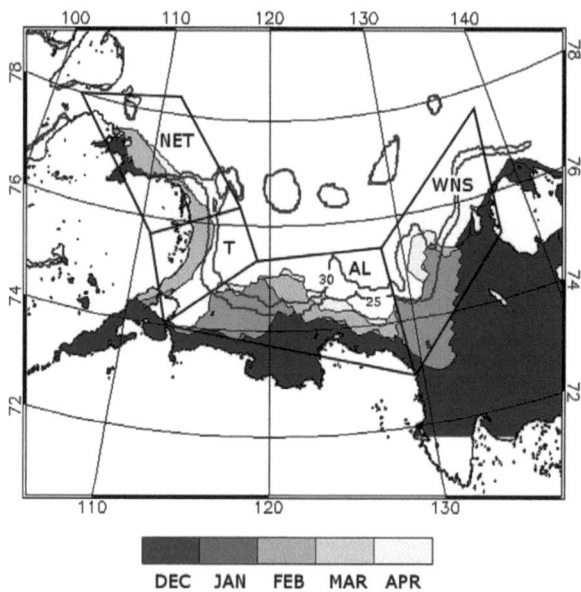

Figure 33: Fast-ice masks derived from MODIS ice-surface temperature for December, 2007 to April, 2008. Contour lines show 25-m and 30-m isobaths. Black bordered areas show the different Laptev Sea polynya subsets: the north-eastern Taimyr (NET) polynya, the Taimyr (T) polynya, the Anabar–Lena (AL) polynya, and the western New Siberian (WNS) polynya. © Figure: Adams et al. [2011], Figure 2.

The second method is based on the monthly TIT composites. The compositing method of the monthly composites differs slightly from the one mentioned in Section 4.6. Ice thickness up to 50 cm is used and the thick-ice pixels are set to 100 cm. This presetting allows artificial ice thickness between 50 and 100 cm when the composites are determined. Fast ice and polynya are easily distinguished using the monthly composites. For better application of the algorithm, the Laptev Sea polynya region is divided into subsets and the fast-ice edge is extracted using a threshold method. This method also requires manual post-processing; partly entailed due to the imperfect cloud mask. The fast-ice masks for the two winter seasons 2007/08 and 2008/09 derived with this method are presented in Figure 34.

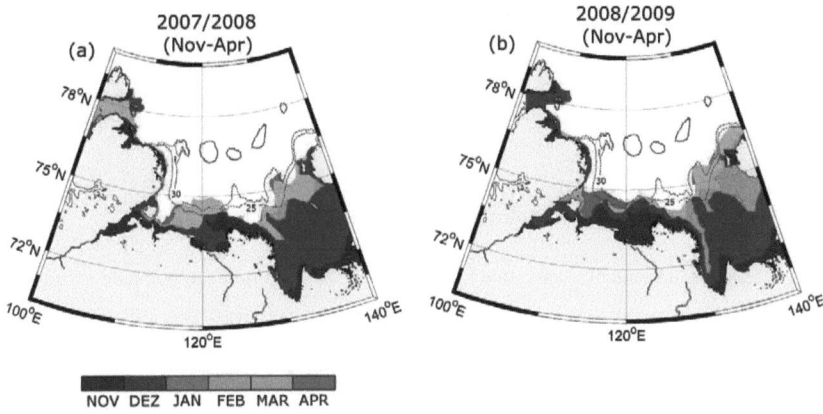

Figure 34: Fast-ice masks derived from monthly MODIS TIT composites for the two winter seasons 2007/08 and 2008/09. Contour lines show 25-m and 30-m isobaths.

While the first fast-ice retrieval is based on one MODIS T_s scene, which thought to be representative of the whole month, the second approach captures the average monthly fast-ice situation. As a result of the inter-diurnal and seasonal variability of the surface temperature it is difficult to create a monthly MODIS T_s composite. Therefore, the fast-ice situation of one MODIS T_s scene per month is selected to represent the fast-ice extent assuming that the conditions do not change considerably over a month. The similarity of the fast-ice masks derived by the two methods supports this assumption. In contrast, the second method provides the monthly averaged fast-ice situation. Fast-ice masks with the MODIS data can be derived from the year 2000 on.

Based on the fast-ice masks, the variability of fast ice can be examined. For the two winter seasons, the fast-ice regions are similar in shape and extent (Figure 34).

For a better overview, an extended study of the fast-ice variability in connection with other oceanographic and atmospheric data is required. As mentioned above, the enormous freshwater discharge into the Laptev Sea has to be examined with respect to the preconditioning of fast-ice evolution (Bareiss and Görgen [2005]).

A precise understanding of the fast-ice formation processes could help to improve sea-ice/ocean models. Up to now, these models are not able to simulate dynamically the formation of fast ice (Wang et al. [2003], König Beatty and Holland [2010]; see Section 5.1). Thus, a fast-ice prescription for these models is necessary. The derived fast-ice masks are appropriate to be implemented into sea-ice/ocean models. The necessity of a high-resolution fast-ice prescription instead of less complex approaches (e.g., using the bathymetry) is described in Section 5.1.

4.7 Intercomparison of various remote sensing data sets

This subsection introduces a case study of 5 January, 2009 in order to compare a variety of remote sensing data sets in terms of the representation of polynyas. The following data sets are used to describe the polynya characteristics:

- MODIS ice-surface temperature;
- MODIS thin-ice thickness;
- MODIS sea-ice concentration;
- AMSR-E polynya signature simulation method;
- AMSR-E sea-ice concentration;
- AMSR-E thin-ice thickness.

4.7.1 Results

Figure 35 presents maps of the Laptev Sea for 5 January, 2009 showing the different data sets. Polynyas occur along the Taimyr Peninsula and along the southern coast of the Laptev Sea Relatively warm MODIS T_s (around -5 °C) shows clearly the polynya in Figure 35a. The daily MODIS thin-ice thickness map presents the TIT distribution derived from MODIS T_s

(Figure 35b). The image indicates the typical polynya structure with open water and very thin ice at the fast-ice edge and increasing ice thickness in the direction of the drift ice.

Applying the 70 % SIC concentrations threshold, MODIS SIC shows smaller polynyas than MODIS TIT (Figure 35b,c). The polynya calculated by MODIS SIC (70 % threshold) is 12,960 km^2 large and the one calculated by MODIS TIT is 29,630 km^2 large (Table 13). The small polynya that is shown by MODIS SIC results from the retrieval method. In order to derive the SIC, a background temperature within a 50 × 50 pixel kernel is calculated (see Section 2.2.3, Drüe and Heinemann [2004]). The background temperature is the median temperature of the kernel and is compared to each individual pixel within the kernel in order to retrieve the SIC. If the difference between the background temperature and the pixel temperature is low in polynya regions, relatively warm T_s result in high ice concentrations. Such a low temperature difference appears when the ice-surface temperatures are similar over a large region (within the kernel). For the case study in January, 2009, T_s is similar over a large area, also within the polynya, resulting in overestimated SIC near the off-shore polynya edge. A possibility to solve this problem is to enlarge the kernel size. However, then the influence of the fast ice and land might yield errors.

In contrast, the case study in December, 2007 shows higher differences between background and pixel temperature in the polynya region. The polynya areas derived from MODIS TIT and MODIS SIC are similar (see Appendix: Figure A6, Table A1).

Along the northern part of the Taimyr Peninsula and the southern coast, the Envisat ASAR image clearly shows a polynya (dark areas) (Figure 35d). The polynya borders are in agreement with the MODIS data sets. However, the ASAR backscatter values do not allow distinguishing between different TIT bands.

Data sets based on AMSR-E 36 GHz or 89 GHz brightness temperatures are presented in Figure 35e-l. PSSM polynya area is calculated according to Willmes et al. [2010]. AMSR-E SIC polynya area is calculated using the 70 % SIC threshold. PSSM POLA agrees well with the AMSR-E SIC POLA. This is in accordance to the polynya studies of Willmes et al. [2010], Willmes et al. [2011] and Adams et al. [2011].

The AMSR-E SIC polynya area is approximately twice as large as MODIS SIC (Table 13). While AMSR-E SIC polynya area agrees well with MODIS TIT, the POLA derived from MODIS SIC seems to be underestimated. The differences in the AMSR-E and MODIS SIC distributions result from the different measurement techniques of the input data (passive microwave versus thermal-infrared) and different retrieval methods (see Section 2.2). While AMSR-E SIC is calculated using the polarization difference between the vertically and horizontally polarized 89 GHz brightness temperatures, the MODIS SIC retrieval is based on T_s differences. Usually, the high-resolution MODIS data is more convenient to derive polynya characteristics but the quality of the method is limited when larger areas show similar temperatures (see above). For this case study, the lower resolution AMSR-E SIC benefits from the wide polynya with large open-water and thin-ice areas. In other cases (e.g., the polynya event on 26 December, 2007) the insensitivity of AMSR-E SIC algorithm to narrow polynyas becomes visible (see Appendix: Figure A6). The MODIS SIC method works well for the case study from December, 2007 because of the high contrast between ice-surface temperatures within the polynya and the background temperature.

The retrieval of AMSR-E thin-ice thickness with the 89 GHz channels (TIT_{89}), the 36 GHz channels (TIT_{36}) and the 36 GHz SIR channels (TIT_{36sir}) is based on Willmes et al. [2010]. Using the correlation of the polarization ratios R_{89}, R_{36} and R_{36sir} and AVHRR TIT, exponential fitting equations are determined to calculate the AMSR-E TIT (see Table 3). The comparison between the polarization ratios reveals the spatial resolution differences between the different frequencies. The typical structure of the polarization ratio within a polynya with high ratios along the fast-ice edge and decreasing ratios going further off-shore is not visible for R_{36}. Instead, the highest ratios are found in the center of the polynya with decreasing ratios in the direction to the fast-ice and to the drift-ice edge. This results from mixed microwave pixels at the polynya edges (Willmes et al. [2010]). The higher resolution of R_{89} (6.25 km × 6.25 km) and R_{36sir} (7.5 km × 7.5 km) show the expected distribution of the polarization ratios within the polynya. However, the R_{36sir} values are higher than the R_{89} values. The distributions of the three polarization ratios confirm the results of Willmes et al. [2010]'s case study.

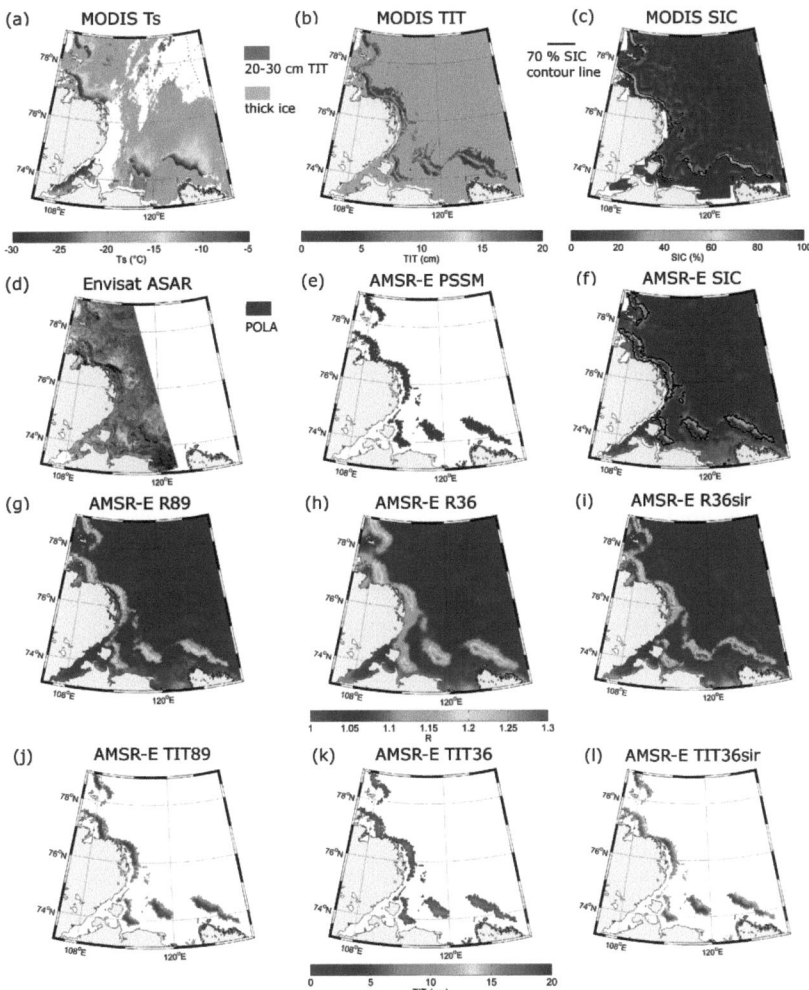

Figure 35: Overview of a various number of satellite data sets for the 5 January, 2009. (a) MODIS ice-surface temperature (T_s) measured at 0550 UTC. White areas are data gaps due to clouds. (b) Daily MODIS thin-ice thickness (TIT) map. (c) Daily MODIS sea-ice concentration (SIC). Black line denotes the 70 % SIC contour line. (d) Envisat ASAR detected at 1308 UTC. (e) Daily AMSR-E PSSM. (f) Daily AMSR-E SIC. Black line denotes the 70 % SIC contour line. (g) Daily AMSR-E polarization ratio of 89 GHz channels (R_{89}). (h) Daily AMSR-E R_{36}. (i) Daily AMSR-E R_{36sir}. (j) Daily AMSR-E thin-ice thickness calculated using 89 GHz channels (TIT_{89}). (k) Daily AMSR-E TIT_{36}. (l) Daily AMSR-E TIT_{36sir}. The TIT distributions in (j) – (l) are masked using the PSSM polynya area. The MODIS data sets and Envisat ASAR data are reprojected to a polar stereographic 1 km × 1 km grid. All AMSR-E data sets are reprojected to a 6.25 km × 6.25 km grid.

For AMSR-E TIT_{89}, TIT_{36} and TIT_{36sir}, only the regions classified as PSSM polynya area are shown (Willmes et al. [2010]). TIT_{89} and TIT_{36sir} show the typical distribution of thin ice within polynyas. In contrast, TIT_{36} is around 10 cm in the polynya center and thinner ice along the drift and fast-ice edge. In accordance to Willmes et al. [2010], TIT_{89} and TIT_{36sir} are valid for polynya monitoring when compared to the higher resolution MODIS and ASAR data sets. One has to be careful using the 89 GHz frequency because of the influence of atmospheric constituents on the measured emissivity. Since the atmospheric influence is negligible for 36 GHz frequency, the enhanced resolution AMSR-E 36 GHz SIR data can be used for operational TIT retrieval (Willmes et al. [2010]). However, the case study for 26 December, 2007 indicates that narrow polynyas are poorly represented by AMSR-E TIT (see Appendix: Figure A6).

Table 13: *The polynya area of 5 January, 2009 is calculated using remote sensing data sets. Polynya area is derived from the thin-ice thickness using the 20 cm contour line as the polynya border. The 70 % threshold is applied for the calculation of polynya area from sea-ice concentration.*

Data set	Polynya area (km^2)
Daily MODIS thin-ice thickness composite	29,630
Daily MODIS sea-ice concentrations	12,960
Daily AMSR-E sea-ice concentrations	35,050
Daily PSSM polynya area	37,780

4.7.2 Conclusions

The results of the two polynya case studies shown in this thesis confirm the results of Willmes et al. [2010]'s case study that the width of the polynya is crucial for the detection of POLA and TIT if AMSR-E data is used (Figure 35; see Appendix: Figure A6). When narrow polynyas occur, the pixels covering the polynya include a mixed signal of open water, thin ice and first-year ice, thus impeding the retrieval of polynya characteristics. In contrast, the daily MODIS composite provides a fine-resolution image of TIT distribution and polynya area. The data gaps occurring in the single MODIS scenes due to clouds can be partly closed by the computing of daily composites. With this method, the average coverage of the Laptev Sea polynya region reaches 70 %.

In comparison to MODIS, Envisat ASAR (150 m × 150 m) provides a higher spatial resolution. However, the daily spatial coverage is limited. Ice is a mixed medium of ice crystals, brine inclusions and air bubbles, occasionally covered with frost flowers or snow. All these components together with the ice's surface roughness have an impact on the electromagnetic properties of the ice layer resulting in highly variable backscatter values (Nghiem and Bertoia [2001]). This impedes the development of an operational TIT algorithm using SAR data. For instance, Kwok et al. [1995] and Dokken et al. [2002] have made initial attempts to derive TIT using SAR data. In order to infer thin-ice thickness from SAR data automatically, a multi-frequency, multi-polarized satellite SAR system is needed to capture the highly variable backscatter features of thin ice (Matsuoka et al. [2001]).

Previous remote sensing studies addressing the ice production use SSM/I or AMSR-E satellite data (Willmes et al. [2011], Tamura and Oshima [2011]). Since narrow polynya events are not captured by coarse-resolution satellite imagery, major errors might occur in the ice production. Hence, a high-resolution data set is required capturing all polynyas and providing a finely resolved thin-ice distribution to calculate accurate ice production. The daily MODIS thin-ice thickness maps, calculated in this thesis, have a spatial resolution of 1 km × 1 km and provide ice thickness up to 20 cm. Thus, the representation of polynya dynamics in this data set might help to derive the polynya ice production more accurately. One shortcoming of this data set is the incomplete coverage.

5 Verification of numerical models using remote sensing data

While global coupled sea-ice ocean models are developed to simulate large-scale sea-ice conditions, FESOM is here analyzed in terms of its performance for polynya simulation. Since the model is not able to simulate fast ice dynamically, the main modification of FESOM is the implementation of fast-ice masks based on MODIS satellite data. Simulated sea-ice fields from different model runs are compared with an emphasis placed on the impact of this prescribed fast-ice mask. This evaluation study is published with modifications in (Adams et al. [2011]).

The first subsection explains the treatment of fast ice in sea-ice models. The second subsection presents an evaluation of FESOM sea-ice concentrations using remote sensing data. The evaluation is focused on the prescribed fast-ice masks.

5.1 Sea-ice/ocean models and the efforts to simulate or prescribe fast ice

The precise reproduction of processes and feedbacks in the Arctic in numerical models is important for the understanding of the Arctic's ocean-sea-ice system. However, recent global coupled sea-ice/ocean models are primarily used to simulate the large-scale sea-ice conditions and ocean processes in the Arctic (Wang et al. [2003], Johnson et al. [2007], Martin and Gerdes [2007]).

Although polynyas have a great impact on properties such as sea-ice concentration, ice growth and ice thickness, as well as water-mass modification and atmospheric circulation patterns (Morales Maqueda et al. [2004], Ebner et al. [2011]), the simulation of these smaller-scale feature is not the main focus of these models. Hence, the realistic simulation of polynya events will be a great challenge for current sea-ice/ocean models. An accurate simulation of the polynya position, as well as its shape and size, is needed for a realistic calculation of ice production in coupled sea-ice/ocean models.

In the Laptev Sea, the fast-ice extent and its variability have a strong influence on the position of the polynyas. Particularly in the South and South East, the polynya location is dependent on the size of the fast-ice area since in these regions the seasonal fast-ice variability is pronounced (see Figure 34).

Currently, sea-ice models are not able to simulate the formation of fast ice (Wang et al. [2003], König Beatty and Holland [2010]). Bathymetry and coastline geometry have already been integrated in sea-ice models, but the shear coefficients typically used are too small for fast ice to remain fixed to the coast during off-shore wind conditions (König Beatty and Holland [2010]). The Laptev Sea polynya is therefore not simulated along the fast-ice edge; rather it is shifted towards the coast. This dislocation of the polynya entails a bias in the simulation of sea-ice concentration, ice growth, ice thickness and ocean winter temperature and salinity distribution (Wang et al. [2003], Rozman [2009]).

To overcome these deficiencies, sea-ice models are modified with respect to their rheology and their numerics to simulate fast-ice features (Lietaer et al. [2008], König Beatty and Holland [2010], Ólason [2011]). The recent approach of Ólason [2011] is successful in simulating fast ice in the Kara Sea. However, further improvements are necessary for an accurate simulation of fast ice with operational coupled sea-ice/ocean models (König Beatty and Holland [2010], Ólason [2011]).

Since sea-ice/ocean models currently are unable to dynamically simulate fast ice, a prescription to simulate its extent and variability is needed. Lieser [2004] produced a fast-ice prescription based on bathymetry. All sea ice in coastal regions having a water depth of less than 30 m is classified as immobile fast ice if the mean ice thickness exceeds one-tenth of the water depth. In terms of model numerics, the respective grid cell is omitted from the grid drift calculations. In summer months, fast ice is reconverted to drift ice to prevent unrealistic ice accumulation in the coastal regions. This simple approach was shown to work well when compared with observations along the Siberian coast (Lieser [2004]).

However, using a relation between bathymetry and ice thickness as an indicator for fast ice is only sufficient for models with a coarse spatial resolution as described in Lieser [2004]. In general, the fast-ice edge follows the 20-30 m isobath depending on time and region (Barber and Hanesiak [2004], Bareiss and Görgen [2005], Mahoney et al. [2007]). As an exception, in some regions the fast-ice edge can also extend over much deeper water, for example, between the islands of the Canadian Arctic Archipelago and on the Russian continental shelves (König Beatty [2007], Mahoney et al. [2007]). This implies that the use of bathymetry for defining the fast-ice edge is only a simplification. It is not sufficient to use the bathymetry as a fast-ice prescription in a fine-resolution sea-ice/ocean model that aims to simulate local processes.

Satellite imagery provides appropriate data from which realistic estimates of the extent and seasonal variability of fast ice can be extracted (Wang et al. [2003]). Thus, high-resolution fast-ice masks derived from MODIS data are implemented in the FESOM model (see Section

4.6.2, Section 3.3.2). In the following subsection, SIC simulations of FESOM are analyzed with respect to the prescribed fast-ice masks.

5.2 Evaluation of simulated sea-ice concentration

5.2.1 Sea-ice concentration data sets

Sea-ice concentration data sets from model configurations with and without fast-ice prescription are evaluated using remote sensing data. Ultimately, three different modeled sea-ice concentration (SIC) data sets of the Laptev Sea are obtained from the coupled sea-ice/ocean model FESOM for the winter season 2007/08. The first SIC data set is obtained from global FESOM model with coarse spatial resolution (FESOM-CR); the second SIC data set is derived from the newly configured fine-scale FESOM-HR model version on a regional grid; and the third SIC data set is derived from the fine-scale FESOM-FI model version which includes a fast-ice parameterization (see Section 3.3.2; see Table 6). All model data sets are available as daily averages. The FESOM-CR data set is available from 1 November, 2007 to 11 May, 2008 (Table 14). FESOM-HR and FESOM-FI ice concentrations are available from 1 April to 11 May, 2008. All model results are interpolated to a common grid of 6.25 km × 6.25 km, which is the resolution of the evaluation data sets.

As evaluation data sets, AMSR-E SIC and PSSM POLA are used. AMSR-E data is used instead of the high-resolution MODIS data because AMSR-E provides a consistent time series (no gaps due to clouds and short wave radiation). Furthermore, the spatial resolution of the model data is similar or coarser than the resolution of AMSR-E data.

Table 14: Basic specifications of the model and remote sensing data sets compared in terms of polynyas. FESOM-CR = global, coarse-resolution model version; FESOM-HR = regional, high-resolution model version; FESOM-FI = regional, high-resolution model version with implemented fast ice.

	FESOM-CR	**FESOM-HR**	**FESOM-FI**	**AMSR-E SIC**	**PSSM**
Spatial resolution (km × km)	~25 × 25	~5 × 5	~5 × 5	6.25 × 6.25	6.25 × 6.25
Temporal resolution	daily mean	daily mean	daily mean	daily mean	daily mean
Period	1 Nov, 2007 – 11 May, 2008	1 April, 2008 – 11 May, 2008	1 April, 2008 – 11 May, 2008	1 Nov, 2007 – 11 May, 2008	1 Nov, 2007 – 11 May, 2008

5.2.2 Evaluation area and variables

For evaluation, the Laptev Sea is divided into different polynya areas according to Bareiss and Görgen [2005] (see Figure 1). From north-west to south-east they are the north-eastern Taimyr (NET) polynya, the Taimyr (T) polynya, the Anabar–Lena (AL) polynya and the western New Siberian (WNS) polynya. WNS and AL polynya represent the eastern Laptev Sea; T and NET represent the western Laptev Sea. A polynya mask that includes all of the four individual polynya areas (called LAP) is additionally used.

Open-water area (OWA) and polynya area (POLA) are calculated from the SIC data sets. OWA gives information about the mean conditions of the modeled sea-ice concentrations in the polynya subsets. This variable is calculated as follows:

$$open-water\,area[m^2] = \sum_{i=1}^{n}(100 - SIC[\%]) * pixel\,area[m^2] \qquad (24)$$

POLA yields information about the location and size of the polynya. Thus, the development of well-formed polynyas in the simulations can be analyzed. Well-formed polynyas are defined as long narrow areas of open water and thin ice. POLA is calculated by means of an ice-concentration threshold. The empirical threshold of 70 % sea-ice concentration yields realistic polynya borders for ice thickness of approximately 15 cm (see Figure 18).

Statistical parameters are calculated for comparison of model and remote sensing data sets. The significance of the correlation coefficient is calculated by the t-test. Correlations above 0.12 (November-May long time series) and above 0.27 (April-May short time series) are significant at the 95 % confidence level.

5.2.3 Open-water area

Figure 36 shows OWA time series calculated from AMSR-E SIC and from the simulated SIC in the polynya subsets from November, 2007 to May, 2008 and April to May, 2008, respectively.

The overestimation of OWA in the FESOM-CR simulations is striking. For the long time series, in the T polynya (western Laptev Sea) the mean of FESOM-CR OWA is approximately 10 times higher than the mean of the AMSR-E data (Table 15a). In the eastern Laptev Sea (WNS, AL), OWA is about two times higher than AMSR-E OWA. The

FESOM-CR OWA correlates weakly with AMSR-E OWA in all of the polynya subsets. Moderate correlations are only found in the NET polynya (r = 0.55).

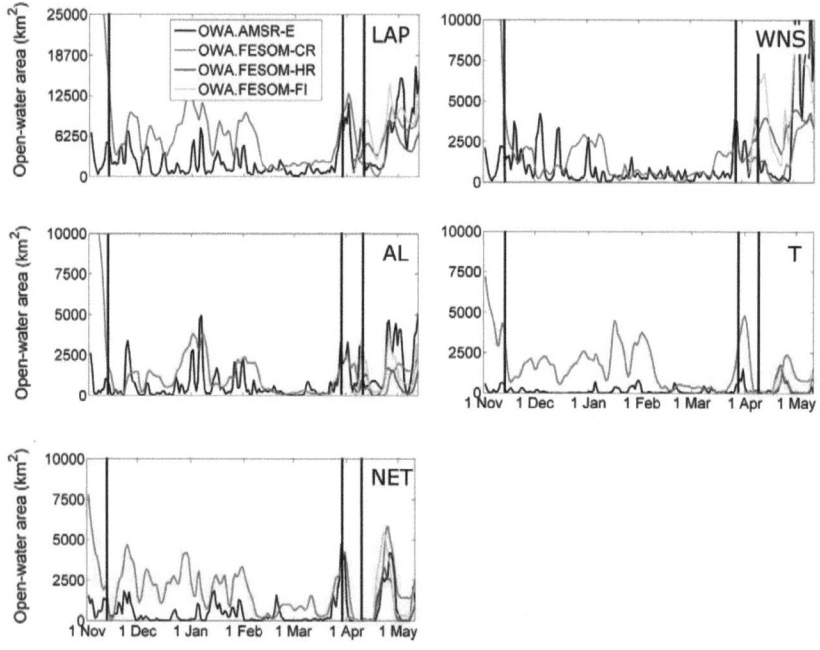

Figure 36: Time series of open-water area (OWA) for the whole polynya system (LAP) and the individual Laptev Sea sub-polynyas (see Figure 1). AMSR-E OWA and FESOM-CR OWA from 1 November, 2007 to 11 May, 2008 as well as FESOM-HR OWA and FESOM-FI OWA from 1 April to 11 May, 2008 are shown. Vertical lines mark polynya events (increase in OWA). © Figure: Adams et al. [2011], Figure 3; modified.

During the six weeks in April and May, 2008, FESOM-CR shows moderate (e.g., T: r = 0.32) to high (e.g., WNS: r = 0.87) correlations with AMSR-E data (Table 15b). Regarding FESOM-HR, the correlations with AMSR-E are lower than the one with FESOM-CR, except in the AL and T polynya (e.g., T: r = 0.63). FESOM-HR OWA is underestimated in the eastern Laptev Sea (WNS, AL) and in agreement with AMSR-E data in the western Laptev Sea (T, NET).

In contrast to the simulations without fast-ice implementation, FESOM-FI OWA is largely consistent with AMSR-E data. OWA in the FESOM-FI simulations is only overestimated during the polynya opening around 8 April, 2008 in the WNS. The correlation between

FESOM-FI and AMSR-E OWA is moderate: between 0.63 in the LAP polynya and 0.74 in the T polynya (r is significant at the 95 % confidence level).

Comparing OWA calculated from the three FESOM model runs, the correlation increases from FESOM-CR to FESOM-HR to FESOM-FI in the AL and T polynya. In all other polynya subsets, the results are ambiguous.

Table 15a: Statistics of the different open-water area (OWA) time series for all polynya subsets for the period from 1 November, 2007 to 11 May, 2008. Mean and standard deviation (SD) of AMSR-E OWA and FESOM-CR OWA are presented. Correlation coefficients (r) are given. Correlations significant at the 95% confidence level (t-test) are in boldface and italics. © Table: Adams et al. [2011], Table 2; modified.

	AMSR-E	FESOM
	Nov–May	
	OWA_{AMSR-E}	$OWA_{FESOM-CR}$
	Sum of all polynya regions (LAP)	
Mean (km^2)	3105	7205
SD (km^2)	3578	7670
r		0.17
	Western New Siberian polynya (WNS)	
Mean (km^2)	1431	2107
SD (km^2)	2298	3479
r		0.12
	Anabar–Lena polynya (AL)	
Mean (km^2)	941	1584
SD (km^2)	1248	2345
r		0.14
	Taimyr polynya (T)	
Mean (km^2)	157	1603
SD (km^2)	243	1391
r		*0.33*
	North-eastern Taimyr polynya (NET)	
Mean (km^2)	576	1911
SD (km^2)	896	1457
r		*0.55*

Polynya openings, characterized here by an increase of OWA, are visible in all data sets, e.g., the polynya events around 27 March, 2008 in the WNS polynya or in late December, 2007 and early January, 2008 in the AL polynya (Figure 36). However, the duration and magnitude of the polynya events are overestimated in FESOM-CR and FESOM-HR. For instance, in the AMSR-E data a polynya opening occurs in the WNS subset at the end of December, 2007 and lasts for a few days. This event is also visible in FESOM-CR but there it lasts for approximately a month. The striking event in the WNS in late April and early May, 2008 is underrepresented in FESOM-CR and FESOM-HR OWA. In contrast to this, FESOM-FI OWA accurately simulates this event.

Table 15b: Statistics of the different open-water area (OWA) time series for all polynya subsets for the period from 1 April to 11 May, 2008. Mean and standard deviation (SD) of AMSR-E OWA, FESOM-CR OWA, FESOM-HR and FESOM-FI are presented. Correlation coefficients (r) are given. Correlations significant at the 95% confidence level (t-test) are in boldface and italics. © Table: Adams et al. [2011], Table 2; modified.

	AMSR-E	FESOM		
		Apr–May		
	OWA_{AMSR-E}	$OWA_{FESOM-CR}$	$OWA_{FESOM-HR}$	$OWA_{FESOM-FI}$
Sum of all polynya regions (LAP)				
Mean (km^2)	6651	5471	5058	7552
SD (km^2)	5244	3550	1779	3276
r		*0.70*	0.46	*0.63*
Western New Siberian polynya (WNS)				
Mean (km^2)	3619	1719	3280	4631
SD (km^2)	3978	1319	862	2580
r		*0.87*	*0.62*	*0.69*
Anabar–Lena polynya (AL)				
Mean (km^2)	1979	1085	706	1270
SD (km^2)	1727	810	561	1084
r		0.49	*0.63*	*0.68*
Taimyr polynya (T)				
Mean (km^2)	207	978	322	384
SD (km^2)	298	1008	482	478
r		0.32	*0.63*	*0.74*
North-eastern Taimyr polynya (NET)				
Mean (km^2)	848	1688	749	1268
SD (km^2)	1267	1683	1026	1670
r		*0.86*	*0.81*	*0.71*

5.2.4 Polynya area

After having compared the mean conditions of the modeled SIC in the polynya subsets, the development of well-formed polynyas in the simulations will be analyzed in the following. POLA is shown as a time series in Figure 37 for the WNS and NET polynya. These two polynya regions can be regarded as representatives of the eastern and western Laptev Sea because the subsets cover large parts of the western and eastern Laptev Sea, respectively. POLA calculated from the two satellite methods (PSSM and SIC threshold) are very consistent. The correlation is around 0.99 (Table 16a). Mean values and standard deviations are very similar for both satellite data sets.

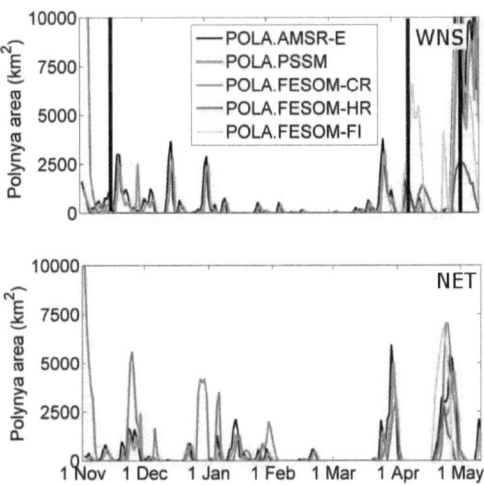

Figure 37: Time series of polynya area (POLA) for winter season 2007/08 in the WNS (eastern Laptev Sea) and NET (western Laptev Sea) polynyas (see Figure 1). AMSR-E POLA, PSSM POLA, FESOM-CR POLA from 1 November, 2007 to 11 May, 2008 as well as FESOM-HR POLA and FESOM-FI POLA from 1 April to 11 May, 2008 are shown. In the WNS polynya, FESOM-CR shows no polynya area after 15 November, 2007 (vertical line). The vertical line at 8 April, 2008 marks an overestimation of FESOM-FI POLA. The vertical line at 30 April, 2008 in WNS marks a polynya event. © Figure: Adams et al. [2011], Figure 4; modified.

The coarse resolution FESOM-CR simulation shows no polynya after 15 November, 2007 in the WNS subset. Accordingly, there is no correlation between FESOM-CR and AMSR-E polynya area (Table 16a). The mean and standard deviation of FESOM-CR POLA in WNS is strongly influenced by the values in the beginning of November. In the western Laptev Sea (NET) polynya activity is better represented by FESOM-CR POLA.

Regarding the period from 1 April to 11 May, 2008 FESOM-CR, shows no polynya in the eastern Laptev Sea (WNS) (Table 16b). In the western Laptev Sea (NET), the correlation between FESOM-CR and AMSR-E is high (r = 0.79), and their mean and standard deviation are similar.

The higher spatial resolution and the reduced diffusivity in FESOM-HR cause polynya activity in the WNS in April and May, 2008. However, FESOM-HR POLA is underestimated in both polynya subsets. The correlation is moderate (e.g., NET: r = 0.68).

Table 16a: Statistics of the different polynya area (POLA) time series of the western New Siberian polynya (WNS) and north-eastern Taimyr polynya (NET) for the period from 1 November, 2007 to 11 May, 2008. Mean and standard deviation (SD) of AMSR-E POLA, PSSM POLA and FESOM-CR POLA is presented. Correlation coefficients (r) are given. Correlations significant at the 95% confidence level (t-test) are in boldface and italics. © Table: Adams et al. [2011], Table 3; modified.

	Remote sensing data		FESOM
	Nov-May		
	$POLA_{AMSR-E}$	$POLA_{PSSM}$	$POLA_{FESOM-CR}$
	Western New Siberian polynya (WNS)		
Mean (km²)	1065	806	521
SD (km²)	2592	2147	3628
r		***0.99***	0.00
	North-eastern Taimyr polynya (NET)		
Mean (km²)	471	364	717
SD (km²)	1028	845	1630
r		***0.98***	***0.43***

Table 16b: Statistics of parameters for the different polynya area (POLA) time series of the western New Siberian polynya (WNS) and north-eastern Taimyr polynya (NET) for the period from 1 April to 11 May, 2008. Mean and standard deviation (SD) of AMSR-E POLA, FESOM-CR POLA, FESOM-HR POLA and FESOM-FI POLA are presented. Correlation coefficients (r) are given. Correlations significant at the 95% confidence level (t-test) are in boldface and italics. © Table: Adams et al. [2011], Table 3; modified.

	AMSR-E	FESOM		
	Apr–May			
	$POLA_{AMSR-E}$	$POLA_{FESOM-CR}$	$POLA_{FESOM-HR}$	$POLA_{FESOM-FI}$
	Western New Siberian polynya (WNS)			
Mean (km²)	3529	0	837	3128
SD (km²)	4722	0	852	3071
r		0.00	***0.59***	***0.60***
	North-eastern Taimyr polynya (NET)			
Mean (km²)	958	821	320	1092
SD (km²)	1556	1939	689	2086
r		***0.79***	***0.68***	***0.56***

The implementation of fast ice results in an increase of POLA in both polynya regions. FESOM-FI POLA shows more consistency with AMSR-E POLA in the NET than in the WNS polynya. In line with the overestimation of FESOM-FI OWA, an overestimation of FESOM-FI POLA in the WNS for the opening in early April, 2008 is found.

Despite the increased spatial resolution of FESOM-HR, the time series of FESOM-HR POLA and their statistical parameters show ambiguous results in terms of polynya simulation. On the other hand, the model results from FESOM-FI show certain improvement. An explanation for this result, that FESOM model improvements do not always yield improved polynya simulations is provided later on.

5.2.5 A case study

Here, the major polynya event that occurred in the eastern Laptev Sea over several days at the end of April, 2008 is examined in more detail. For this event the daily average SIC distributions of AMSR-E, FESOM-CR, FESOM-HR and FESOM-FI on 29 April, 2008 are shown in Figure 38 as an example.

FESOM-CR simulations without fast ice yield no polynya area in the WNS region but very homogeneous ice coverage with some reduced SIC in the eastern Laptev Sea (Figure 38b). There, the SIC reaches about 79 %. Examination of the FESOM-CR simulations during the winter season shows that well-formed polynyas are not visible over the entire period (Figure 37). Instead, FESOM-CR simulations feature a SIC reduced to 60-80 % over a larger area in the polynya regions during the events.

The high-resolution model FESOM-HR simulates coastal polynyas with open-water areas along the coastline and an increasing ice coverage within the polynya going farther off-shore (Figure 38c).

With a prescribed fast-ice cover, FESOM-FI SIC shows improved results with distinctive polynyas simulated along the fast-ice edge (Figure 38d). The comparison with AMSR-E SIC fields shows a large degree of consistency with the polynya location but also a small shift towards the coastline, as seen in the difference plot (Figure 38g).

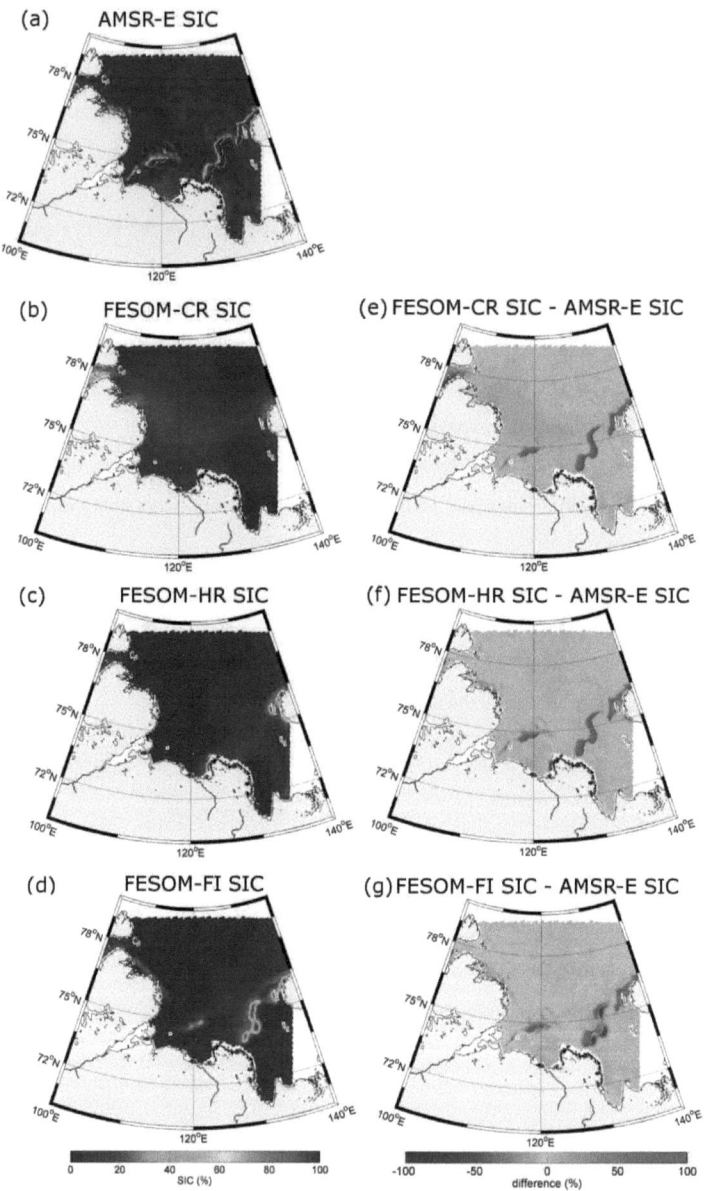

Figure 38: Maps of sea-ice concentration (SIC) on 29 April, 2008 derived from: (a) AMSR-E; (b) FESOM-CR; (c) FESOM-HR and (d) FESOM-FI. Plots showing the difference that result when AMSR-E SIC is subtracted from (e) FESOM-CR SIC; (f) FESOM-HR SIC and (g) FESOM-FI SIC. © Figure: Adams et al. [2011], Figure 6; modified.

5.2.6 Summary and discussion

Table 17 summarizes the general findings of the three simulations. The comparison between the results of OWA and POLA shows a general overestimation of FESOM-CR OWA, while FESOM-CR POLA is underestimated. The reason for this is that a large open-water area does not correspond to a formation of a well-formed polynya with strongly reduced SIC. In fact, the large FESOM-CR OWA results from slightly reduced sea-ice concentrations over large areas. Regarding the polynya position, polynyas are not present or are found in the wrong location in the coarse-resolution model when run without fast ice.

Table 17: Comparison of sea-ice concentration (SIC) and polynya features simulated by the three model versions of FESOM. © Table: Adams et al. [2011], Table 4; modified.

Feature	FESOM with fast ice, coarse resolution (FESOM-CR)	FESOM without fast ice, high resolution (FESOM-HR)	FESOM with fast ice, high resolution (FESOM-FI)
SIC in general	generally underestimated	except for polynya regions: good agreement	good agreement
Polynya location	dislocation, located at the coastline	dislocation, located at the coastline	realistic location at the fast-ice edge
Polynya shape	large areas with slightly reduced SIC, no well-formed polynya	depends on coastline, no agreement with AMSR-E	accurate
SIC in polynya	no well-formed polynya, slightly reduced SIC over a large area	small-scale gradients, lowest SIC are reproduced	small-scale gradients, lowest SIC are reproduced

FESOM-HR also simulates the polynyas at the wrong position but shows a high SIC gradient within the polynyas. In regions (e.g., NET and WNS) where in reality the polynyas are located near to the coast of the mainland or an island due to a small fast-ice area, the agreement with AMSR-E is better, as it is in regions with a large fast-ice extension (e.g., AL). For simulations that include fast ice (FESOM-FI), the polynyas are simulated at the correct positions. The small displacement of the polynya between AMSR-E and FESOM-FI data could result from the difference between the two remote sensing data sets. For the prescribed fast-ice edge high-resolution MODIS data (1 km pixel size) were used, while the comparison used coarser resolution AMSR-E data (6.25 km pixel size).

The comparison of polynya areas from satellite data using two different algorithms (SIC threshold and PSSM) shows a very strong correlation between the mentioned two data sets. Despite the fact that SIC errors can be up to 10 %, or higher at lower concentrations (Andersen et al. [2007], Spreen et al. [2008]), and that the PSSM has a slight tendency to underestimate POLA (Willmes et al. [2010]), both data sets appear to give reliable estimates of the true polynya conditions.

A previous comparison of a sea-ice/ocean model equivalent to FESOM-CR with passive microwave satellite products from Kauker et al. [2003] shows general agreement between both data sets on a large scale in terms of the long-term mean state and the interseasonal variability of the simulated SIC. This is supported by Wang et al. [2003], who demonstrated that current sea-ice/ocean models put more emphasis on the representation of large-scale sea-ice extent and concentration in the Arctic and Antarctic. Consequently, the distribution of sea-ice concentration is homogenous with only gentle gradients in FESOM-CR simulations that have a coarser spatial resolution and exclude fast ice.

While this smoothing effect seems appropriate to match the large-scale SIC distribution, it also leads to a blurring of the polynya signature (Wang et al. [2003]) such that the simulated SIC is overestimated at the polynya location and underestimated in the drift-ice and fast-ice areas. The spatial smoothing of SIC could also induce a smoothing in time which would explain the overestimated duration of polynya activity.

There are two basic reasons for the smoothing of the SIC in space and time. The first is the insufficient spatial resolution of FESOM-CR simulations (grid spacing of 25 km) as well as an insufficient spatial and temporal resolution of the forcing data (approximately 200 km, 6-hourly). A second reason is that FESOM-CR simulations were realized with a relatively large horizontal diffusivity (2000 m^2 s^{-1}), which clearly produces very smooth SIC fields and a deceptive weakening of polynya signatures.

Regarding the SIC distribution within polynyas, AMSR-E data show zero to low sea-ice concentrations at the fast-ice edge; going farther off-shore the SIC increases (Figure 38a). This graduated distribution is not reproduced by FESOM-CR. Rather, the sea-ice concentrations are homogeneously distributed within the polynya regions at medium concentrations. The minima of SIC are not shown (Figure 38b). This implies that the simulated polynyas consist mostly of thin ice with only little open water. Consequently, the heat flux between ocean and atmosphere in FESOM-CR would be underestimated. According to Ebner et al. [2011] an ice layer of a few centimeters thickness reduces considerably the heat loss to the atmosphere. Hence, ice-production values calculated by FESOM-CR might have major errors.

Rollenhagen et al. [2009] point out a general underestimation of ice concentrations (i.e., an overestimation of OWA) from FESOM-CR in comparison with satellite data. This is consistent with the results of this evaluation study. The coarse spatial resolution has an impact on the underestimation of SIC. Polynyas are sub-grid scale phenomena for coarse-resolution models. To mimic the effect of polynyas, the model simulates low concentrations in a broader area. Furthermore, the daily wind forcing data (i.e., temporally smoothed wind fields) cannot resolve short-term events that may be crucial for a realistic description of polynya formation.

In FESOM-HR the spatial resolution has been increased (approximately 5 km grid spacing compared to 25 km in FESOM-CR), and the forcing data have a higher spatial resolution (approximately 40 km compared to 200 km in FESOM-CR). Additionally, the horizontal diffusion has been reduced (100 m^2 s^{-1} compared to 2000 m^2 s^{-1} in FESOM-CR). These improvements avoid a spurious smoothing of SIC. However, the smoothing of SIC in FESOM-CR results in a masking of the model errors in terms of small-scale effects. In the fine-scale FESOM-HR simulations, these errors become apparent. For instance, uncertainties in the momentum fluxes (transfer coefficient and wind forcing) might lead to an overestimation of OWA and POLA, e.g., in the WNS polynya around 8 April, 2008. Improvements of the turbulence closure scheme, e.g., a stability-dependent transfer coefficient, are of rising importance when the resolution is increased. However, the gradual increase of SIC within the polynya showing very low SIC at the coast to levels of approximately 70 % farther seawards in the polynya is realistically reproduced.

The introduction of fast ice is more pronounced in the eastern Laptev Sea due to the greater extent of fast ice in this region. In the western part of the Laptev Sea, along the Taimyr Peninsula, the extent of the fast ice is only 10 to 20 km due to the very steep slope of the seafloor (Reimnitz et al. [1995]). In particular, FESOM-CR OWA is seriously overestimated in the western Laptev Sea in comparison to the eastern Laptev Sea. This means that in FESOM-CR the polynya regions are also correctly positioned in the western Laptev Sea when no fast ice is involved.

The evaluation study concludes that FESOM-CR (coarse spatial resolution, no fast-ice parameterization) is able to reduce the sea-ice concentrations during polynya events (Figure 36). Regarding the SIC fields, the reason for the reduction of SIC becomes clear. As mentioned above, the smoothing of SIC is important for this effect, not the simulation of realistic polynyas (Figure 38b). When polynyas are simulated, they are not located in the expected regions. Furthermore, the improved model run FESOM-HR is not able to solve this problem. This implies that modifications in the numerics of the model (with our current knowledge) would not lead to improvements concerning the location of polynyas. At this point, it becomes obvious that fast ice has to be included. Hence, a kind of fast-ice parameterization is essential for a realistic simulation of the small-scale processes in the Laptev Sea.

The improvement of the model results due to the implementation of fast ice can be clearly seen in the SIC fields (Figure 38d,g). FESOM-FI SIC show correctly simulated polynyas along the edge of the fast ice.

The analysis demonstrates that a fast-ice prescription in sea-ice/ocean models is necessary but not sufficient for a realistic sea-ice concentration simulation. The required model improvements are more complex. Therefore, on a regional scale the following improvements for FESOM are proposed:

(1) improving the turbulence closure scheme (e.g., the stability-dependent transfer coefficient);

(2) optimizing the momentum fluxes (transfer coefficient and wind forcing);

(3) using finer spatial and temporal resolution forcing data.

These improvements should help to optimize the simulations of FESOM with respect to small-scale processes and features.

6 Concluding remarks

6.1 Conclusions

This thesis presents a methodological improvement for the monitoring of thin sea ice, a quality analysis of the variables required for this method and the resulting thin-ice thickness (TIT).

The thermal-infrared thin-ice thickness retrieval of Yu and Lindsay [2003] is successfully improved with respect to the parameterizations of the turbulent heat and the long-wave radiation fluxes. These surface-energy fluxes are updated to state-of-the-art parameterizations for the atmospheric boundary layer (e.g., including atmospheric stratification). The analysis of the heat transfer coefficient (C_H) shows that a constant C_H, as used by Yu and Lindsay [2003] and Drucker and Martin [2003], is applicable as a first approach. However, the variability of C_H indicates that the improved algorithm results in a more precise calculation of turbulent heat fluxes and TIT.

The sensitivity analysis of the algorithm indicates that the strong temperature differences between the surface and the atmosphere in areas of open water and very thin ice yield a masking effect with respect to the errors of the atmospheric variables. For TIT less than 10 cm, the impact of the uncertainty in the input variables is therefore almost negligible. However, since TIT is calculated also for thicker ice, a feasible atmospheric data set should be used for the TIT calculation.

The analysis of the input variables demonstrates that ice-surface temperature T_s and 2-m air temperature T_a are the essential variables for the TIT retrieval (Wang et al. [2010]). Uncertainties in these variables yield a pronounced impact on the accuracy of the calculated TIT. Regarding the atmospheric variables, the commonly used NCEP T_a (e.g., Drucker and Martin [2003]) is compared to COSMO T_a. COSMO has a finer spatial resolution (5 km grid space instead of 200 km) and includes the polynya in contrast to NCEP.

Air temperature data sets that do not capture polynyas (as a boundary condition) underestimate T_a above polynyas, which results in underestimated TIT. However, the warm bias in NCEP T_a partly compensates the temperature underestimation above polynyas.

In COSMO, the initial sea-ice field is prescribed by AMSR-E SIC data. This means that the size and position of COSMO's polynyas are identical to the polynyas presented by AMSR-E. Taking into account the representation of polynyas and the good agreement of COSMO data with in-situ measurements (Ernsdorf et al. [2011]), the usage of the COSMO data should result in more accurate TIT results. However, the differences between AMSR-E and MODIS data (e.g., AMSR-E polynyas are often narrower than MODIS polynyas) result in inconsistencies between MODIS T_s and COSMO T_a affecting the thin-ice thickness calculation (TIT jumps, underestimation of TIT).

The analysis result revealed a larger TIT error based on an underestimation of T_a in contrast to an overestimation of T_a (note here the warm bias of NCEP). Due to this NCEP data will be used for a long term TIT calculation. Moreover, this data set is globally available and allows the algorithm to be applied to other polar shelf regions.

The statistical sensitivity analysis shows a mean absolute error of ±4.7 cm for thin-ice thickness up to 20 cm. The error increases strongly when the ice is thicker than 20 cm. Although previous studies, e.g., Yu and Lindsay [2003] and Drucker and Martin [2003], used the thermal-infrared thin-ice thickness up to 50 cm, this thesis proposes that for subsequent studies exclusively thin-ice thickness below 20 cm should be used.

Thin-ice thickness fields in the Laptev Sea are calculated using MODIS ice-surface temperatures and NCEP atmospheric variables, for the winter seasons 2007/08 and 2008/09. The data set has a spatial resolution of 1 km × 1 km. Due to the restriction of the algorithm to cloud-free and nighttime pixels, the single MODIS thin-ice thickness scenes show incomplete coverage. To overcome this drawback, daily thin-ice thickness composites are calculated. These maps cover on average 70 % of the Laptev Sea polynya. The daily TIT maps are valuable as a verification data set, as input data for the retrieval of monthly fast-ice composites and as an assimilation data set for sea-ice ocean models (see below).

As an overall conclusion, it is pointed out that the results of this thesis push the monitoring of polynya dynamics forward. The presented results contribute substantially to a better understanding of methodological issues concerning the thermal-infrared TIT retrieval and provide for the first time high-resolution maps showing the TIT distribution within the Laptev Sea. The daily TIT composites enable a computation of a more precise ice production values.

6.2 Outlook

One important aim of the TIT monitoring is to improve estimates of ice production within polynya systems. Providing daily MODIS TIT composites is the first step towards increasing the quality of ice production results since the requirement of high-resolution TIT distributions are substantial for ice-production estimations (Willmes et al. [2011]). Based on this, future work can concern:

(1) the further improvement of the TIT methodology;
(2) improvement of the cloud identification in the MODIS data;
(3) the allocation of a full coverage TIT data set based on a remote sensing – model assimilation method;
(4) the calculation of ice-production values.

In terms of the improvement of the TIT algorithm, a future perspective could be a closer coupling with atmospheric model data. For example, the radiation fluxes from the COSMO model could be used instead of simple empirical parameterizations.

Regarding the MODIS ice-surface temperatures the problems with unidentified clouds and their impact on the TIT retrieval are addressed. Thin-ice thickness artifacts emerge in the TIT maps due to the warm cloud temperature. In particular, for nighttime scenes the quality of the cloud mask is lower because the visible MODIS channels cannot be used for cloud detection. Further studies are required to improve cloud identification in MODIS data. For instance, a cloud masking algorithm based on a time series of MODIS images could be applied (Hall et al. [2004], Lyapustin et al. [2008]). Improved cloud detection would substantially improve the TIT calculation.

Referring to polynya representation the analysis of the atmospheric model temperature illustrates the differences between the model data and the MODIS data. Whereas NCEP 2-m air temperature (T_a) does not represent the polynyas in general, COSMO T_a predominantly captures polynyas underestimated in width. A bias correction based on MODIS T_s could be applied to adjust COSMO and NCEP T_a respectively. Such a correction method might be a possibility to compensate the shortcomings of the model air temperature and could yield an increased quality of the MODIS thin-ice thickness data set.

The daily MODIS thin-ice thickness composites developed here still lack complete coverage. Closing the gaps with other remote sensing data sets (e.g., AMSR-E data) is problematic due to the inconsistencies between the data sets. A possible solution to obtain gap-free TIT maps is the assimilation of MODIS TIT into a sea-ice model.

The global coupled sea-ice/ocean model FESOM is improved with respect to a fast-ice parameterization based on MODIS fast-ice masks, its numerics and the forcing data. The improved regional model version is sufficient to simulate the polynya dynamics in the Laptev

Sea; however, the model is not always able to provide realistic thin-ice thickness maps. Therefore, the daily MODIS thin-ice thickness composites can be assimilated with optimal interpolation into FESOM (D. Schröder, personal communication, 2012). The aim of the assimilation method is to provide a consistent time series of thin-ice thickness maps with a spatial resolution of 5 km × 5 km. FESOM can also be used to calculate the ice production.

Future research should focus on methodological improvements to increase the quality of the MODIS thin-ice thickness maps, and further improvement and analysis of the remote sensing – model assimilation method. The challenge of this combined approach is the calculation of polynya ice production using high-resolution thin-ice thickness data. The included MODIS TIT distribution allows a more accurate computation of the heat loss to the atmosphere, and hence, more accurate estimation of ice production values in comparison to previous studies.

Bibliography

Ackerman, S. A., R. E. Holz, R. Frey, E. W. Eloranta, B. C. Maddux, and M. McGill (2008), Cloud Detection with MODIS. Part II: Validation, *J. Atmos. Oceanic Technol.* 25(7), 1073–1086, DOI: 10.1175/2007JTECHA1053.1.

Ackerman, S. A., K. I. Strabala, W. P. Menzel, R. A. Frey, C. C. Moeller, and L. E. Gumley (1998), Discriminating clear sky from clouds with MODIS, *J. Geophys. Res.* 103(D24), 32,141–32,157, DOI: 10.1029/1998JD200032.

Adams, S., S. Willmes, D. Schröder, G. Heinemann, M. Bauer, and T. Krumpen (2012), Improvement and sensitivity analysis of thermal thin-ice retrievals, *IEEE Trans. Geosci. Remote Sensing*, in revision.

Adams, S., S. Willmes, G. Heinemann, P. Rozman, R. Timmermann, and D. Schröder (2011), Evaluation of simulated sea-ice concentrations from sea-ice/ocean models using satellite data and polynya classification methods, *Polar Res.* 30(7124), pp 17, DOI: 10.3402/polar.v30i0.7124.

Alexandrov, V. Y., T. Martin, J. Kolatschek, H. Eicken, M. Kreyscher, and A. P. Makshtas (2000), Sea ice circulation in the Laptev Sea and ice export to the Arctic Ocean: Results from satellite remote sensing and numerical modeling, *J. Geophys. Res.* 105(C7), 17,143–17,159, DOI: 10.1029/2000JC900029.

Andersen, S., R. Tonboe, L. Kaleschke, G. Heygster, and L. T. Pedersen (2007), Intercomparison of passive microwave sea ice concentration retrievals over the high-concentration Arctic sea ice, *J. Geophys. Res.* 112(C08004), pp 18, DOI: 10.1029/2006JC003543.

Andreas, E. L. (1987), A theory for the scalar roughness and the scalar transfer coefficients over snow and sea ice, *Boundary-Layer Meteorol.* 38(1-2), 159–184, DOI: 10.1007/BF00121562.

Antonova, S. (2011), Spatial and temporal variability of the fast ice in the Russian Arctic, *Master thesis*, State University of St. Petersburg, St. Petersburg, Russia and University of Hamburg, Hamburg, Germany, pp 37.

Arctic Climate Assessment (ACIA) (2004), Impacts of a Warming Arctic, *Scientific Report*, Cambridge Univ. Press, New York, USA, 139 pp.

Barber, D., and R. Massom (2007), The role of sea ice in Arctic and Antarctic polynyas, *In: Polynyas: Windows to the world. Edited by W.O. Smith Jr. and D.G. Barber*, Oceanography Series 74, Elsevier, Amsterdam, The Netherlands, pp 54.

Barber, D. G., and J. M. Hanesiak (2004), Meteorological forcing of sea ice concentrations in the southern Beaufort Sea over the period 1979 to 2000, *J. Geophys. Res.* 109(C06014), pp 16, DOI: 10.1029/2003JC002027.

Bareiss, J., and K. Görgen (2005), Spatial and temporal variability of sea ice in the Laptev Sea: Analyses and review of satellite passive-microwave data and model results, 1979 to 2002, *Global Planet Change* 48(1-3), 28–54, DOI: 10.1016/j.gloplacha.2004.12.004.

Bareiss, J. (2003), Süßwassereintrag und Festeis in der ostsibirischen Arktis - Ergebnisse aus Boden- und Satellitenbeobachtungen sowie Sensitivitätsstudien mit einem thermodynamischen Festeismodell, *Ph.D. thesis*, Universtity of Trier, Trier, Germany, pp 170.

Barnes, W., T. Pagano, and V. Salomonson (1998), Prelaunch characteristics of the Moderate Resolution Imaging Spectroradiometer (MODIS) on EOS-AM1, *IEEE Trans. Geosci. Remote Sensing 36*(4), 1088–1100, DOI: 10.1109/36.700993.

Brutsaert, W. (1975), On a derivable formula for long-wave radiation from clear skies, *Water Resources Research 11*(5), 742–744, DOI: 10.1029/WR011i005p00742.

Cavalieri, D., T. Markus, and J. Comiso (2004), AMSR-E/Aqua Daily L3 12.5 km brightness temperature, sea ice concentration, & snow depth polar grids V002, November 2007 – April 2009, *Digital media*. *Updated daily*, National Snow and Ice Data Center, Boulder, USA.

Cavalieri, D., C. Parkinson, P. Gloersen, and H. J. Zwally (1996), Sea Ice Concentrations from Nimbus-7 SMMR and DMSP SSM/I-SSMIS Passive Microwave Data, *Digital media*. *Updated yearly*, National Snow and Ice Data Center, Boulder, USA.

Cavalieri, D. J., P. Gloersen, and W. J. Campbell (1984), Determination of Sea Ice Parameters With the NIMBUS 7 SMMR, *J. Geophys. Res. 89*(D4), 5355–5369, DOI: 10.1029/JD089iD04p05355.

Ciappa, A., L. Pietranera, and G. Budillon (2012), Observations of the Terra Nova Bay (Antarctica) polynya by MODIS ice surface temperature imagery from 2005 to 2010, *Remote Sens. Environ. 119*(0), 158–172, DOI: 10.1016/j.rse.2011.12.017.

Comiso, J. C., C. L. Parkinson, R. Gersten, and L. Stock (2008), Accelerated decline in the Arctic sea ice cover, *Geophys. Res. Lett. 35*(L01703), pp 6. DOI: 10.1029/2007GL031972.

Comiso, J. C., and K. Steffen (2001), Studies of Antarctic sea ice concentrations from satellite data and their applications, *J. Geophys. Res. 106*(C12), 31,361–31,385, DOI: 10.1029/2001JC000823.

Comiso, J. C., and A. L. Gordon (1998), Interannual variability in summer sea ice minimum, coastal polynyas and bottom water formation in the Weddell Sea, *In: Antarctic Sea Ice: Physical Processes, Interaction, and Variability*. Edited by M. O. Jeffries, Antarct. Res. Ser. 74, AGU, Washington, D. C., USA, 293–315.

Dethleff, D., P. Loewe, and E. Kleine (1998), The Laptev Sea flaw lead - detailed investigation on ice formation and export during 1991/1992 winter season, *Cold Reg. Sci. Technol. 27*(3), 225–243, DOI: 10.1016/S0165-232X(98)00005-6.

Dmitrenko, I. A., S. A. Kirillov, L. B. Tremblay, D. Bauch, and S. Willmes (2009), Sea-ice production over the Laptev Sea shelf inferred from historical summer-to-winter hydrographic observations of 1960s–1990s, *Geophys. Res. Lett. 36*(L13605), pp 4, DOI: 10.1029/2009GL038775.

Dmitrenko, I., K. Tyshko, S. Kirillov, H. Eicken, J. Hölemann, and H. Kassens (2005), Impact of flaw polynyas on the hydrography of the Laptev Sea, *Global Planet Change* *48*(1-3), 9–27, DOI: 10.1016/j.gloplacha.2004.12.016.

Dmitrenko, I. A., V. A. Gribanov, D. L. Volkov, H. Kassens, and H. Eiken (1999), Impact of river discharge on the fast ice extension in the Russian Arctic shelf area, *Proceedings of the 15th International Conference on Port and Ocean Engineering under Arctic Conditions (POAC99), Tech. Rep. 1*, Helsinki, Finland, 23-27 August.

Dokken, S. T., P. Winsor, T. Markus, J. Askne, and G. Björk (2002), ERS SAR characterization of coastal polynyas in the Arctic and comparison with SSM/I and numerical model investigations, *Remote Sens. Environ. 80*(2), 321–335, DOI: 10.1016/S0034-4257(01)00313-3.

Drucker, R., S. Martin, and R. Kwok (2011), Sea ice production and export from coastal polynyas in the Weddell and Ross Seas, *Geophys. Res. Lett. 38*(L17502), pp 4, DOI: 10.1029/2011GL048668.

Drucker, R., and S. Martin (2003), Observations of ice thickness and frazil ice in the St. Lawrence Island polynya from satellite imagery, upward looking sonar, and salinity/temperature moorings, *J. Geophys. Res. 108*(C5), pp 18. DOI: 10.1029/2001JC001213.

Drüe, C., and G. Heinmann (2005), Accuracy assessment of sea-ice concentrations from MODIS using in-situ measurements, *Remote Sens. Environ.*(95), 139–149, DOI: 10.1016/j.rse.2004.12.004.

Drüe, C., and G. Heinemann (2004), High-resolution maps of the sea-ice concentration from MODIS satellite data, *Geophys. Res. Lett. 31*(L20403), pp 5, DOI: 10.1029/2004GL020808.

Early, D., and D. Long (2001), Image reconstruction and enhanced resolution imaging from irregular samples, *IEEE Trans. Geosci. Remote Sensing 39*(2), 291–302, DOI: 10.1109/36.905237.

Ebner, L., D. Schröder, and G. Heinemann (2011), Impact of the Laptev Sea flaw polynyas on the atmospheric boundary layer and ice production using idealized mesoscale simulations., *Polar Res. 30*(7210), pp 16, DOI: 10.3402/polar.v30i0.7210.

Efimova, N. A. (1961), On methods of calculating monthly values of net long wave radition, *Meteorologiya i Gidrologiya 10*, 28–33.

Eicken, H., J. Kolatschek, J. Freitag, F. Lindemann, H. Kassens, and I. Dmitrenko (2000), A key source area and constraints on entrainment for basin-scale sediment transport by Arctic sea ice, *Geophys. Res. Lett. 27*(13), 1919–1922. DOI: 10.1029/1999GL011132.

Ernsdorf, T., D. Schröder, S. Adams, G. Heinemann, R. Timmermann, and S. Danilov (2011), Impact of atmospheric forcing data on simulations of the Laptev Sea polynya dynamics using the sea-ice ocean model FESOM, *J. Geophys. Res. 116*(C12038), pp 18, DOI: 10.1029/2010JC006725.

Fissel, D. B., and C. L. Tang (1991), Response of sea ice drift to wind forcing on the northeastern Newfoundland shelf, *J. Geophys. Res. 96*(C10), 18,397-18,409, DOI: 10.1029/91JC01841.

Fraser, A. D., R. A. Massom, and K. J. Michael (2010), Generation of high-resolution East Antarctic landfast sea-ice maps from cloud-free MODIS satellite composite imagery, *Remote Sens. Environ. 114*(12), 2888–2896, DOI: 10.1016/j.rse.2010.07.006.

Fraser, A., R. Massom, and K. Michael (2009), A Method for Compositing Polar MODIS Satellite Images to Remove Cloud Cover for Landfast Sea-Ice Detection, *IEEE Trans. Geosci. Remote Sensing 47*(9), 3272–3282, DOI: 10.1109/TGRS.2009.2019726.

Frey, R. A., S. A. Ackerman, Y. Liu, K. I. Strabala, H. Zhang, J. R. Key, and X. Wang (2008), Cloud Detection with MODIS. Part I: Improvements in the MODIS Cloud Mask for Collection 5, *J. Atmos. Oceanic Technol. 25*(7), 1057–1072, DOI: 10.1175/2008JTECHA1052.1.

Friedrich, A. (2010), Untersuchung der Polynja-Dynamik im Weddellmeer, Antarktis, mit Mikrowellenfernerkundung, *Diploma thesis*, Department of Environmental Meteorology, Universtity of Trier, Trier, Germany, pp 79.

Giles, K. A., S. W. Laxon, and A. L. Ridout (2008), Circumpolar thinning of Arctic sea ice following the 2007 record ice extent minimum, *Geophys. Res. Lett. 35*(L22502), pp 4, DOI: 10.1029/2008GL035710.

Gloersen, P., D. J. Cavalieri, A. T. C. Chang, T. T. Wilheit, W. J. Campbell, O. M. Johannessen, K. B. Katsaros, K. F. Kunzi, D. B. Ross, D. Staelin, E. P. L. Windsor, F. T. Barath, P. Gudmandsen, E. Langham, and R. O. Ramseier (1984), A Summary of Results From the First NIMBUS 7 SMMR Observations, *J. Geophys. Res. 89*(D4), 5335–5344, DOI: 10.1029/JD089iD04p05335.

Graversen, R. R., and M. Wang (2009), Polar amplification in a coupled climate model with locked albedo, *Clim Dyn 33*(5), 629–643, DOI: 10.1007/s00382-009-0535-6.

Groves, J. E., and W. J. Stringer (1991), The Use of AVHRR Thermal Infrared Imagery to Determine Sea Ice Thickness within the Chukchi Polynya, *Arctic 44,* 130–139.

Gudmandsen, P. (2005), Lincoln Sea and Nares Strait, *Proccedings of the 2004 Envisat and ERS Symposium (ESA SP-572),* Salzburg, Austria, 6-10 September.

Gudmandsen, P. (2000), A Remote Sensing Study of Lincoln Sea, *Proccedings ERS-Envisat Symposium, Paper No. 409 (CD-ROM),* ESA Publication Division, Gothenburg, Sweden, pp 8.

Guenther, B., X. Xiong, V. Salomonson, W. Barnes, and J. Young (2002), On-orbit performance of the Earth Observing System Moderate Resolution Imaging Spectroradiometer; first year of data, *Remote Sens. Environ. 83*(1-2), 16–30, DOI: 10.1016/S0034-4257(02)00097-4.

Haas, C., S. Hendricks, and M. Doble (2006), Comparison of the sea-ice thickness distribution in the Lincoln Sea and adjacent Arctic Ocean in 2004 and 2005, *Ann. Glaciol. 44*(1), 247–252, DOI: 10.3189/172756406781811781.

Hall, D. K., G. A. Riggs, and V. V. Salomonson (2007), MODIS/Terra and MODIS/Aqua Sea Ice Extent 5-min L2 Swath 1km V005, 1 December 2007 - 30 April 2008 and 1 December 2008 - 30 April 2009, *Digital media.* Updated daily, National Snow and Ice Data Center, Boulder, USA.

Hall, D. K., J. R. Key, K. A. Case, G. A. Riggs, and D. J. Cavalieri (2004), Sea ice surface temperature product from MODIS, *IEEE Trans. Geosci. Remote Sensing 42*(5), 1076–1087, DOI: 10.1109/TGRS.2004.825587.

Heil, P. (1999), Sea-ice growth, drift and deformation off east Antarctica, *Ph.D. thesis*, Antarctic CRC and Institute of Antarctic and Southern Ocean Studies, University of Tasmania, Hobart, Australia, pp 243.

Heitronics Infrarot Messtechnik GmbH (2012), Infrared Radiation Pyrometer KT15 II, *Operating Instructions*, Wiesbaden, Germany, pp 74.

Hollinger, J. P., R. Lo, and G. Poe (1987), Special Sensor Microwave/Imager, *User's Guide*, Naval Research Laboratory, Washington, D.C., USA.

Hollinger, J., J. Peirce, and G. Poe (1990), SSM/I instrument evaluation, *IEEE Trans. Geosci. Remote Sensing 28*(5), 781–790, DOI: 10.1109/36.58964.

Hughes, N., J. Wilkinson, and P. Wadhamns (2011), Multi-satellite sensor analysis of fast-ice development in the Norske er Ice Barrier, northeast Greenland, *Ann. Glaciol. 52*(57), 151–160.

Hunke, E. C., and J. K. Dukowicz (1997), An elastic-viscous-plastic model for sea ice dynamics., *J. Phys. Oceanogr.*(27), 1849–1868.

Ishikawa, T., J. Ukita, K. I. Ohshima, M. Wakatsuchi, T. Yamanouchi, and N. Ono (1996), Coastal polynyas off East Queen Maud Land observed from NOAA AVHRR data, *J. Oceanogr. 52*(3), 389–398, DOI: 10.1007/BF02235932.

Jin, X., D. Barber, and T. Papakyriakou (2006), A new clear-sky downward longwave radiative flux parameterization for Arctic areas based on rawinsonde data, *J. Geophys. Res. 111*(D24104), pp 7, DOI: 10.1029/2005JD007039.

Johannessen, O. M. (2005), *Remote sensing of sea ice in the Northern Sea Route. Studies and applications*, Springer, Berlin, Germany, pp 472.

Johnson, M., S. Gaffigan, E. Hunke, and R. Gerdes (2007), A comparison of Arctic Ocean sea ice concentration among the coordinated AOMIP model experiments, *J. Geophys. Res. 112*(C04S11), pp 16, DOI: 10.1029/2006JC003690.

Kaleschke, L. (2003), Fernerkundung des Meereises mit passiven und aktiven Mikrowellensensoren, *Ph.D. thesis*, Fachbereich für Physik und Elektrotechnik, University of Bremen, Bremen, Germany, pp 204.

Kaleschke, L., C. Lüpkes, T. Vihma, J. Haarpaintner, A. Bochert, J. Hartmann, and G. Heygster (2001), SSM/I Sea Ice Remote Sensing for Mesoscale Ocean-Atmosphere Interaction Analysis, *Can. J. Remote Sens. 27*(5), 526–537.

Kanamitsu, M., W. Ebisuzaki, J. Woollen, S.-K. Yang, J. J. Hnilo, M. Fiorino, and G. L. Potter (2002), NCEP-DOE AMIP-II Reanalysis (R-2), *Bull. Amer. Meteor. Soc. 83,* 1631–1643, DOI: 10.1175/BAMS-83-11-1631.

Kauker, F., R. Gerdes, M. J. Karcher, C. Köberle, and J. Lieser (2003), Variability of Arctic and North Atlantic sea ice: A combined analysis of model results and observations from 1978 to 2001, *J. Geophys. Res. 108*(C6, 3182), pp 20, 10.1029/2002JC001573.

Kern, S. (2009), Wintertime Antarctic coastal polynya area: 1992-2008, *Geophys. Res. Lett. 36*(L14501), pp 5, DOI: 10.1029/2009/GL038062.

Kern, S., G. Spreen, L. Kaleschke, S. De la Rosa, and G. Heygster (2007), Polynya Signature Simulation Method polynya area in comparison to AMSR-E 89 GHz sea-ice concentrations in the Ross Sea and off the Adélie Coast, Antarctica, for 2002–05: first results, *Ann. Glaciol. 46*(1), 409–418, DOI: 10.3189/172756407782871585.

Key, J. R., J. B. Collins, C. Fowler, and R. S. Stone (1997), High-Latitude Surface Temperature Estimates from Thermal Satellite Data., *Remote Sens. Environ. 61,* 302–309, DOI: 10.1016/S0034-4257(97)89497-7.

Kneizys, F. X., E. P. Shettle, L. W. Abreu, J. H. Chetwynd, G. W. Anderson, W. O. Gallery, J. E. A. Selby, and S. A. Clough (1988), User's Guide to LOWTRAN 7, *Tech. Rep. AFGL-TR-88-0177 (NTIS AD A206773),* Air Force Geophys. Lab., Hanscom Air Force Base, Bedfort, USA.

König Beatty, C., and D. M. Holland (2010), Modeling Landfast Sea Ice by Adding Tensile Strength, *J. Phys. Oceanogr. 40*(1), 185–198, DOI: 10.1175/2009JPO4105.1.

König Beatty, C. (2007), Arctic Landfast Sea Ice, *Ph.D. thesis,* Center for Atmosphere Ocean Science, Department of Mathematics, New York University, New York, USA, pp 110.

Kozo, T. (1991), The hybrid polynya at the northern end of Nares Strait, *Geophys. Res. Lett. 18,* 2059–2062, DOI: 10.1029/91GL02574.

Krell, A., C. Ummenhofer, G. Kattner, A. Naumov, D. Evans, G. S. Dieckmann, and D. N. Thomas (2003), The biology and chemistry of land fast ice in the White Sea, Russia? A comparison of winter and spring conditions, *Polar Biol. 26*(11), 707–719, DOI: 10.1007/s00300-003-0543-7.

Kuhn, P. M., L. P. Sterns, and R. O. Ramseier (1975), Airborne infrared imagery of arctic sea ice thickness, *NOAA, Tech. Report ERL 331 - APCL 34,* U.S. Department of Commerce, NOAA, Environmental Research Laboratories, Boulder, USA.

Kwok, R., L. Toudal Pedersen, P. Gudmandsen, and S. S. Pang (2010), Large sea ice outflow into the Nares Strait in 2007, *Geophys. Res. Lett. 37*(L03502), pp 6, DOI: 10.1029/2009GL041872.

Kwok, R., and D. A. Rothrock (2009), Decline in Arctic sea ice thickness from submarine and ICESat records: 1958–2008, *Geophys. Res. Lett. 36*(L15501), DOI: 10.1029/2009GL039035.

Kwok, R., J. Comiso, S. Martin, and R. Drucker (2007), Ross Sea polynyas: Response of ice concentration retrievals to large areas of thin ice, *J. Geophys. Res. 112*(C12012), pp 13, DOI: 10.1029/2006JC003967.

Kwok, R. (2005), Variability of Nares Strait ice flux, *Geophys. Res. Lett. 32*(L24502), pp 4, DOI: 10.1029/2005GL024768.

Kwok, R., S. Nghiem, S. Yueh, and D. Huynh (1995), Retrieval of thin ice thickness from multifrequency polarimetric SAR data, *Remote Sens. Environ. 51*(3), 361–374, DOI: 10.1016/0034-4257(94)00017-H.

Lantuit, H., and W. Pollard (2008), Fifty years of coastal erosion and retrogressive thaw slump activity on Herschel Island, southern Beaufort Sea, Yukon Territory, Canada, *Geomorphology 95*(1-2), 84–102, DOI: 10.1016/j.geomorph.2006.07.040.

Launiainen, J., and T. Vihma (1990), Derivation of turbulent surface fluxes — An iterative flux-profile method allowing arbitrary observing heights, *Environ. Softw. 5*(3), 113–124, DOI: 10.1016/0266-9838(90)90021-W.

Lieser, J. L. (2004), A numerical model for short-term sea ice forecasting in the Arctic, Ein numerisches Modell zur Meereisvorhersage in der Arktis, *Ph.D. thesis*, Fachbereich Physik / Elektrotechnik, University of Bremen, Bremen, Germany, pp 106.

Lietaer, O., T. Fichefet, and V. Legat (2008), The effects of resolving the Canadian Arctic Archipelago in a finite element sea ice model, *Ocean Model. 24*(3-4), 140–152, DOI: 10.1016/j.ocemod.2008.06.002.

Long, D., and J. Stroeve (2011), Enhanced-Resolution SSM/I and AMSR-E Daily Polar Brightness Temperatures, *Digital media*, National Snow and Ice Data Center, Boulder, USA.

Lu, J., and M. Cai (2009), Seasonality of polar surface warming amplification in climate simulations, *Geophys. Res. Lett. 36*(L16704), pp 6, DOI: 10.1029/2009GL040133.

Lyapustin, A., Y. Wang, and R. Frey (2008), An automatic cloud mask algorithm based on time series of MODIS measurements, *J. Geophys. Res. 113*(D16207), pp 15, DOI: 10.1029/2007JD009641.

Lythe, M., A. Hauser, and Wendler G. (1999), Classifcation of sea ice types in the Ross Sea, Antarctica from SAR and AVHRR imagery, *Int. J. Remote Sensing 20*(15-16), 3073–3085, DOI: 10.1080/014311699211624.

Mahoney, A., H. Eicken, and L. Shapiro (2007), How fast is landfast sea ice? A study of the attachment and detachment of nearshore ice at Barrow, Alaska, *Cold Reg. Sci. Technol. 47*(3), 233–255, DOI: 10.1016/j.coldregions.2006.09.005.

Mahoney, A., H. Eicken, A. Graves, L. Shapiro, and P. Cotter (2004), Landfast sea ice extent and variability in the Alskan Arctic derived from SAR imagery, *IEEE Inter. Geosci. Remote Sensing Symposium Proceedings 3*, 2146–2149, DOI: 10.1109/IGARSS.2004.1370783.

Majewski, D., D. Liermann, P. Prohl, B. Ritter, M. Buchhold, T. Hanisch, G. Paul, W. Wergen, and J. Baumgardner (2002), The operational global icosahedral-hexagonal grid point model GME: Description and high resolution tests, *Mon. Wea. Rev. 130*, 319–388, DOI: 10.1175/1520-0493(2002)130<0319:TOGIHG>2.0.CO;2.

Markus, T., D. Cavalieri, and A. Ivanoff (2011), Algorithm Theoretical Basis Document for the AMSR-E Sea Ice Algorithm, Revised December 2011, *Tech. Report,* Goddard Space Flight Center, Landover, USA.

Markus, T., D. Cavalieri, and A. Ivanoff (2008), AMSR-E Algorithm Theoretical Basis Document: Sea Ice Products, *Tech. Report,* Goddard Space Flight Center, Greenbelt, USA.

Markus, T., C. Kottmeier, and E. Fahrbach (1998), Ice formation in coastal polynyas in the Weddell Sea and their impact on oceanic salinity, *In: Antarctic Sea Ice: Physical Processes, Interaction,and Variability. Edited by M. O. Jeffries,* AGU, Washington, D. C., USA, 293-315.

Markus, T., and B. A. Burns (1995), A method to estimate subpixel-scale coastal polynyas with satellite passive microwave data, *J. Geophys. Res. 100*(C3), 4473–4487.

Marsland, S. J. (2004), Modeling water mass formation in the Mertz Glacier Polynya and Adélie Depression, East Antarctica, *J. Geophys. Res. 109*(C11003), pp 18, DOI: 10.1029/2004JC002441.

Martin, S. (2005), Improvements in the estimates of ice thickness and production in the Chukchi Sea polynyas derived from AMSR-E, *Geophys. Res. Lett. 32*(L05505), pp 4, DOI: 10.1029/2004GL022013.

Martin, S. (2004), Estimation of the thin ice thickness and heat flux for the Chukchi Sea Alaskan coast polynya from Special Sensor Microwave/Imager data, 1990–2001, *J. Geophys. Res. 109*(C10012), pp 15, DOI: 10.1029/2004JC002428.

Martin, S. (2001), Polynyas, *In: Encyclopedia of Ocean Sciences. Edited by J. H. Steele, K. K. Turekian and S. A. Thorpe,* Academic Press, London, UK, 2241–2247, DOI: 10.1006/rwos.2001.0007.

Martin, T., and R. Gerdes (2007), Sea ice drift variability in Arctic Ocean Model Intercomparison Project models and observations, *J. Geophys. Res. 112*(C04S10), pp 13, DOI: 10.1029/2006JC003617.

Massom, R. A., K. Hill, C. Barbraud, N. Adams, A. Ancel, L. Emmerson, and M. J. Pook (2009), Fast ice distribution in Adélie Land, East Antarctica: Interannual variability and implications for emperor penguins Aptenodytes forsteri, *Mar. Ecol. Prog. Ser. 374,* 243–257, DOI: 10.3354/meps07734.

Massom, R. A., K. L. Hill, V. I. Lytle, A. P. Worby, M. J. Paget, and I. Allison (2001), Effects of regional fast-ice and iceberg distributions on the behaviour of the Mertz Glacier polynya, East Antarctica, *Ann. Glaciol. 33,* 391–398.

Massom, R. A., P. T. Harris, K. J. Michael, and M. J. Potter (1998), The distribution and formative processes of latent-heat polynyas in East Antarctica, *Ann. Glaciol. 27,* 420–426.

Matsuoka, T., S. Uratsuka, M. Satake, T. Kobayashi, A. Nadai, T. Umehara, H. Maeno, H. Wakabayashi, K. Nakamura, and F. Nishio (2001), CRL/NASDA airborne SAR (Pi-SAR) observations of sea ice in the Sea of Okhotsk, *Ann. Glaciol. 33*(1), 115–119, DOI: 10.3189/172756401781818734.

Maykut, G. A. (1986), The surface heat and mass balance, *In: Geophysics of Sea Ice. Edited by N. Untersteiner,* Plenum Press, NATO Advanced Science Institutes Series B, Physics, New York, USA, 395–463.

Maykut, G. A., and N. Untersteiner (1971), Some results from a time-dependent thermodynamic model of sea ice, *J. Geophys. Res. 76*, 1550–1575, DOI: 10.1029/JC076i006p01550.

Meier, W., F. Fetterer, K. Knowles, M. Savoie, and M. J. Brodzik (2006), Sea Ice Concentrations from Nimbus-7 SMMR and DMSP SSM/I-SSMIS Passive Microwave Data, *Digital media. Updated quarterly,* National Snow and Ice Data Center, Boulder, USA.

Morales Maqueda, M. A., A. J. Willmot, and N. R. T. Biggs (2004), Polynya dynamics: A review of observations and modeling., *Rev. Geophys. 42*(RG1004), pp 37, DOI: 10.1029/2002RG000116.

Naoki, K., J. Ukita, F. Nishio, M. Nakayama, J. C. Comiso, and A. Gasiewski (2008), Thin sea ice thickness as inferred from passive microwave and in situ observations, *J. Geophys. Res. 113*(C02S16), pp 11, DOI: 10.1029/2007JC004270.

Nghiem, S., and C. Bertoia (2001), Multi-polarization C-band SAR signatures of arctic sea ice, *IEEE Inter. Geosci. Remote Sensing Symposium Proceedings 3,* 1243–1245, DOI: 10.1109/IGARSS.2001.976806.

Nishihama, M., Wolfe R. E., Solomon D., Patt F. S., Blanchette J., Fleig A. J., and Masuoka E. (1997), MODIS Level 1A Earth Location Algorithm Theoretical Basis Document Version 3.0, SDST-092, *Tech. Report,* Lab. Terrestrial Phys. NASA Goddard Space Flight Center, Greenbelt, USA.

Ólason, E. Ö. (2011), Dynamical modelling of Kara Sea land-fast ice, *Ph.D. thesis,* Department Geowissenschaften, University of Hamburg, Hamburg, Germany, pp 231.

Parkinson, C. L., J. C. Comiso, H. J. Zwally, D. J. Cavalieri, P. Gloerson, and W. J. Campbell (1987), Arctic Sea-Ice, 1973-1976: Satellite Passive-Microwave Observations, *Tech. Report NASA SP-489,* National Aereonautics and Space Administration, Washington, D.C., USA.

Parkinson, C. L., and W. M. Washington (1979), A large scale numerical model of sea ice, *J. Geophys. Res. 84,* 311–337, DOI: 10.1029/JC084iC01p00311.

Project 'System Laptev Sea' (2011), Eurasische Schelfmeere im Umbruch - Ozeanische Fronten und Polynyasystme in der Laptev See II, *Interim report*, pp 252.

Rees, W. G. (1993), Infrared emissivities of Arctic land cover types, *Int. J. Remote Sensing 14*(5), 1013–1017, DOI: 10.1080/01431169308904392.

Reimnitz, E., H. Eicken, and T. Martin (1995), Multiyear fast ice along the Taymyr Peninsula, Siberia., *Arctic*(48, 4), 359–377.

Renfrew, I. A., G. W. K. Moore, P. S. Guest, and K. Bumke (2002), A Comparison of Surface Layer and Surface Turbulent Flux Observations over the Labrador Sea with ECMWF Analyses and NCEP Reanalyses, *J. Phys. Oceanogr. 32*(2), 383–400, DOI: 10.1175/1520-0485(2002)032<0383:ACOSLA>2.0.CO;2.

Riggs, G. A., D. K. Hall, and V. V. Salomonson, MODIS Sea Ice (2012), *User's Guide*, Online available: http:\\modis-snow-ice.gsfc.nasa.gov/siugkc.html.

Rigor, I., and R. Colony (1997), Sea-ice production and transport of pollutants in the Laptev Sea, 1979–1993, *Sci. Total Environ. 202*(1-3), 89–110, DOI: 10.1016/S0048-9697(97)00107-1.

Rollenhagen, K., R. Timmermann, T. Janjić, J. Schröter, and S. Danilov (2009), Assimilation of sea ice motion in a finite-element sea ice model, *J. Geophys. Res. 114*(C05007), pp 14, DOI: 10.1029/2008JC005067.

Rozman, P. (2009), The Role of the Laptev Sea Fast Ice in an Arctic Ocean – Sea Ice Coupled Model., *Master thesis*, Faculty of Geography and Geoecolgy, Saint Petersburg State University, Saint Petersburg, Russia, pp 57.

Schättler, U., G. Doms, and C. Schraff (2009), A description of the non-hydrostatic regional COSMO-Model, Part VII., *User's Guide*, Deutscher Wetterdienst, Offenbach, Germany, pp 147.

Schröder, D., G. Heinemann, and S. Willmes (2011), Implementation of a thermodynamic sea ice module in the NWP model COSMO and its impact on simulations for the Laptev Sea area in the Siberian Arctic, *Polar Res. 30*(5974), pp 18, DOI: 10.3402/polar.v30i0.5971.

Schröder, D., T. Vihma, A. Kerber, and B. Brümmer (2003), On the parameterization of turbulent surface fluxes over heterogeneous sea ice surfaces, *J. Geophys. Res. 108*(C6, 3195), pp 12, DOI: 10.1029/2002JC001385.

Screen, J. A., and I. Simmonds (2010), The central role of diminishing sea ice in recent Arctic temperature amplification, *Nature 464*(7293), 1334–1337, DOI: 10.1038/nature09051.

Selyuzhenok, V. (2011), Validation of satellite-based landfast ice mapping, *Master thesis*, State University of St. Petersburg, St. Petersburg, Russia and University of Hamburg, Hamburg, Germany, pp 32.

Serreze, M. C., A. P. Barrett, J. C. Stroeve, D. N. Kindig, and M. M. Holland (2009), The emergence of surface-based Arctic amplification, *The Cryosphere 3*(1), 11–19, DOI: 10.5194/tc-3-11-2009.

Serreze, M. C., and J. A. Francis (2006), The Arctic Amplification Debate, *Climatic Change 76*(3-4), 241–264, DOI: 10.1007/s10584-005-9017-y.

Smith, S. D., R. D. Muench, and C. H. Pease (1990), Polynyas and Leads: An Overview of Physical Processes and Environment, *J. Geophys. Res. 95*(C6), 9461–9479, DOI: 10.1029/JC095iC06p09461.

Spinhirne, J. D. (2005), Antarctica cloud cover for October 2003 from GLAS satellite lidar profiling, *Geophys. Res. Lett. 32*(L22S05), pp 4, DOI: 10.1029/2005GL023782.

Spinhirne, J. D., S. P. Palm, D. L. Hlavka, W. D. Hart, and A. Mahesh (2004), Global and polar cloud cover from the geoscience laser altimeter system, observations and implication, *In: AGU Fall Meeting Abstracts*.

Spreen, G., L. Kaleschke, and G. Heygster (2008), Sea ice remote sensing using AMSR-E 89-GHz channels, *J. Geophys. Res. 113*(C02S03), pp 14, DOI: 10.1029/2005JC003384.

Spreen, G., L. Kaleschke, and G. Heygster (2005), Operational sea ice remote sensing with AMSR-E 89 GHz channels, *IEEE Inter. Geosci. Remote Sensing Symposium Proceedings 6*, 4033–4036, DOI: 10.1109/IGARSS.2005.1525799.

Steppeler, J., G. Doms, U. Schättler, H. W. Bitzer, A. Cassmann, U. Damrath, and G. Gregoric (2003), Meso-gamma scale forecasts using the nonhydrostatic model LM, *Meteorol. Atmos. Phys. 82*, 75–96, DOI: 10.1109/IGARSS.2005.1525799.

Tamura, T., and K. I. Oshima (2011), Mapping of sea ice production in the Arctic coastal polynyas, *J. Geophys. Res. 116*(C07030), pp 20, DOI: 10.1029/2010JC006586.

Tamura, T., K. I. Ohshima, and S. Nihashi (2008), Mapping of sea ice production for Antarctic coastal polynyas, *Geophys. Res. Lett. 35*(L07606), pp 5, DOI: 10.1029/2007GL032903.

Tamura, T., K. I. Ohshima, T. Markus, D. J. Cavalieri, S. Nihashi, and N. Hirasawa (2007), Estimation of Thin Ice Thickness and Detection of Fast Ice from SSM/I Data in the Antarctic Ocean, *J. Atmos. Oceanic Technol. 24*(10), 1757–1772, DOI: 10.1175/JTECH2113.1.

Tanaka, S., K. Suzuki, T. Yamanouchi, and S. Kawaguchi (1985), Atmospheric effects against the surface temperature measurement by AVHRR in the polar region, *Mem. Natl. Polar Res., Spec. Issue 39*, 80–86.

Ukita, J., T. Kawamura, N. Tanaka, T. Toyota, and M. Wakatsuchi (2000), Physical and stable isotopic properties and growth processes of sea ice collected in the southern Sea of Okhotsk, *J. Geophys. Res. 105*(C9), 22,083–22,093. DOI: 10.1029/1999JC000013.

Van Woert, M. L. (1999), Wintertime dynamics of the Terra Nova Bay polynya, *J. Geophys. Res. 104*(C4), 7753–7769, DOI: 10.1029/1999JC900003.

Wang, J., R. Kwok, F. J. Saucier, J. Hutchings, M. Ikeda, W. Hibler III, J. Haapala, M. D. Coon, H. E. M. Meier, H. Eicken, N. Tanaka, D. Prentki, and W. Johnson (2003), Working Toward Improved Small-scale Sea Ice-Ocean Modeling in the Arctic Seas, *Eos Trans. AGU 84*(34), 325–336.

Wang, X., J. R. Key, and Y. Liu (2010), A thermodynamic model for estimating sea and lake ice thickness with optical satellite data, *J. Geophys. Res. 115*(C12035), pp 14, DOI: 10.1029/2009JC005857.

Willmes, S., S. Adams, D. Schröder, and G. Heinemann (2011), Spatio-temporal variability of polynya dynamics and ice production in the Laptev Sea between the winters of 1979/80 and 2007/08, *Polar Res. 30*(5971), pp 16, DOI: 10.3402/polar.v30i0.5971.

Willmes, S., T. Krumpen, S. Adams, L. Rabenstein, C. Haas, J. Hölemann, S. Hendricks, and G. Heinemann (2010), Cross-validation of polynya monitoring methods from multisensor satellite and airborne data: a case study for the Laptev Sea, *Can. J. Remote Sens. 36*, 196–210, DOI: 10.5589/m10-012.

Winsor, P., and G. Björk (2000), Polynya activity in the Arctic Ocean from 1958 to 1997, *J. Geophys. Res. 105*(C4), 8789–8803, DOI: 10.1029/1999JC900305.

Wolfe, R. E., M. Nishihama, A. J. Fleig, J. A. Kuyper, D. P. Roy, J. C. Storey, and F. S. Patt (2002), Achieving sub-pixel geolocation accuracy in support of MODIS land science, *Remote Sens. Environ. 83*(1-2), 31–49, DOI: 10.1016/S0034-4257(02)00085-8.

World Meteorological Organization (1990), WMO sea-ice nomenclature. Terminology, codes and illustrated glossary, *Tech. Report 259*, Secretariat of the World Meteorological Organization, Geneva, Switzerland.

Xiong, X., K. Stamnes, and D. Lubin (2002), Surface Albedo over the Arctic Ocean Derived from AVHRR and its Validation with SHEBA Data, *J. Appl. Meteor. Climatol. 41*(4), 413–425, DOI: 10.1175/1520-0450(2002)041<0413:SAOTAO>2.0.CO;2.

Yu, Y., and R. W. Lindsay (2003), Comparison of thin ice thickness distributions derived from RADARSAT Geophysical Processor System and advanced very high resolution radiometer data sets, *J. Geophys. Res. 108*(C12, 3387), pp 11, DOI: 10.1029/2002JC001319.

Yu, Y., D. A. Rothrock, and J. Zhang (2001), Thin ice impacts on surface salt flux and ice strength: Inferences from advanced very high resolution radiometer, *J. Geophys. Res. 106*(C7), 13,975–13,988.

Yu, Y., and D. A. Rothrock (1996), Thin ice thickness from satellite thermal imagery, *J. Geophys. Res. 101*(C11), 25,753–25,766. DOI: 10.1029/96JC02242.

List of Symbols

Symbol	Name	(Value) / Unit
C_E	Transfer coefficient for evaporation	-
C_H	Transfer coefficient for heat	-
C_{HN}	Heat transfer coefficient for neutral stratification	-
c_p	Specific heat of air	1.0035 J g^{-1} K^{-1}
E_0	Latent heat flux	W m^{-2}
e_a	Atmospheric water vapor pressure	hPa
H	Horizontal polarization	-
H_0	Sensible heat flux	W m^{-2}
h_i	Ice thickness	m
k_i	Conductivity of pure ice	2.03 W K^{-1} m^{-1}
L	Latent heat of vaporization	2500 J g$_{-1}$
L*	Obukhov length	m
L↑	Upward long-wave radiation flux	W m^{-2}
L↓	Downward long-wave radiation flux	W m^{-2}
L_f	Latent heat of freezing	0.334 J kg^{-1}
p	Sea-level pressure	hPa
P_0	Polarization difference including the atmospheric influence for sea-ice concentration = 0 %	-
P_1	Polarization difference including the atmospheric influence for sea-ice concentration = 100 %	-
P_d	Polarization difference	-
PR	Polarization ratio	-
$P_{s,i}$	Polarization difference for ice	-
$P_{s,w}$	Polarization difference for water	-
Q_0	Net radiation balance	W m^{-2}
q_a	Atmospheric specific humidity	g kg^{-1}
Q_a	Net energy flux to the atmosphere	W m^{-2}
Q_I	Conductive heat flux through the ice	W m^{-2}
q_s	Surface specific humidity	g kg^{-1}
R	Simple polarization ratio	-
Re	Roughness Reynolds number	-
RH	Relative humidity	%
SIC	Sea-ice concentration	%
T_a	2-m air temperature	°C
T_B	Brightness temperature	K
T_{bg}	Background temperature	°C
T_f	Freezing temperature of sea water	-1.8 °C
T_{max}	Maximum temperature	°C

T_s	Ice-surface temperature	°C
T_{th}	Temperature threshold	°C
u_*	Friction velocity	m s-1
U_{10m}	10-m wind speed	m s^{-1}
U_{2m}	2-m wind speed	m s-1
V	Vertical polarization	-
z_0	Roughness length for momentum	m
z_q	Roughness length for humidity	m
z_t	Roughness length for temperature	m
ε_a	Atmospheric emission coefficient	-
ε_s	Surface emission coefficient	-
ζ	Monin-Obukhov stability	-
θ	Sensor scan angle	°
$\overline{\theta}$	Mean potential temperature	K
θ_a	Potential temperature in the atmosphere	K
θ_s	Potential temperature at surface	K
ρ_a	Air density	kg m^{-3}
ρ_i	Density of sea ice	910 kg m^{-3}
σ	Stefan-Boltzmann constant	5.67×10^{-8} W m^{-2} K^{-4}

List of Abbreviations

Abbreviation	Name
ACIA	Arctic Climate Assessment
ADCP	Acoustic Doppler Current Profiler
AL	Anabar-Lena polynya
AMSR-2	Advanced Microwave Scanning Radiometer – 2
AMSR-E	Advanced Microwave Scanning Radiometer – Earth Observation System
APECS	Association of Polar Early Career Scientists
ASAR	Advanced Synthetic Aperture Radar
ASI	ARTIST sea-ice algorithm
AVHRR	Advanced Very High Resolution Radiometer
AWI	Alfred Wegener Institute for Polar and Marine Research
AWS	Automatic Weather Stations
BMBF	German Ministry for Education and Research
CDT	Conductivity Temperature Depth Sensor
COSMO	Consortium for Small-scale Modeling
DAAD	German Academic Exchange Service
DMSP	Defense Meteorological Satellite Program
DOE	Department of Energy
DWD	German Weather Service
ECMWF	European Centre for Medium-Range Weather Forecasts
EF61	Efimova [1961]
Envisat	Environmental Satellite
ESA	European Space Agency
ESMR	Electrically Scanning Microwave Radiometer
FESOM	Finite Element Sea-Ice Ocean Model
FESOM-CR	Finite Element Sea-Ice Ocean Model, coarse resolution, global
FESOM-FI	Finite Element Sea-Ice Ocean Model, high resolution, regional and fast-ice prescription
FESOM-HR	Finite Element Sea-Ice Ocean Model, high resolution, regional
FI	Fast Ice
FOV	Field Of View
GCOM-W1	Global Change Observation Mission 1st - Water
GME	Global Model Extended
GPS	Global Positioning System
JI06	Jin et al. [2006]
KT15	KT 15 II pyrometer
LOWTRAN	Low Resolution Radiative Transfer Model
MOD/MYD021KM	Level 1B Calibrated Radiances - 1 km
MOD/MYD03	Geolocation - 1 km

MOD/MYD29	MODIS Sea Ice Extent 5-Min L2 Swath 1km
MOD/MYD35_L2	MODIS Level 2 Cloud Mask and Spectral Test Results
MODIS	Moderate Resolution Imaging Spectroradiometer
NASA	US National Aeronautics and Space Administration
NCAR	US National Center for Atmospheric Research
NCEP	US National Center for Environmental Prediction
NET	North-eastern Taimyr polynya
NOAA	US National Oceanic and Atmospheric Administration
NSIDC	US National Snow and Ice Data Center
POLA	Polynya Area
POTOWA	Potential Open-water Algorithm
PSSM	Polynya Signature Simulation Method
R_{36}	Polarization ratio of vertical and horizontal polarized AMSR-E 36 GHz channels
R_{36sir}	Polarization ratio of vertical and horizontal polarized enhanced-resolution AMSR-E 36 GHz channels
R_{89}	Polarization ratio of vertical and horizontal polarized AMSR-E 89 GHz channels
RMSE	Root Mean Square Error
SAR	Synthetic Aperture Radar
SIC	Sea-ice Concentrations
SIR	Scatterometer Image Reconstruction
SMMR	Scanning Multichannel Microwave Radiometer
SSM/I	Special Sensor Microwave Imager
T	Taimyr polynya
TDXV	Transdrift XV
TIT	Thin-ice Thickness
TIT_{36}	Thin-ice Thickness calculated using AMSR-E 36 GHz channels
TIT_{36sir}	Thin-ice Thickness calculated using the enhanced-resolution AMSR-E 36 GHz channels
TIT_{89}	Thin-ice Thickness calculated using AMSR-E 89 GHz channels
WNS	Western New Siberian polynya

Appendix

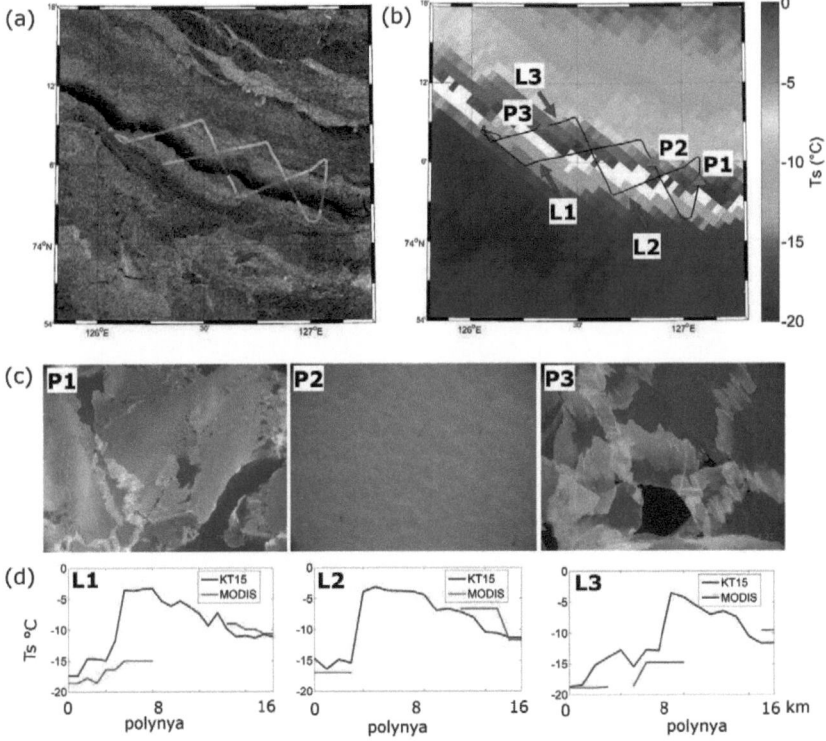

Figure A1: (a) Envisat ASAR image from 26 March, 2009 0232 UTC for illustration of the polynya conditions. Color-graded transect was measured at 26 March, 2009 around 0830 UTC and shows KT15 ice-surface temperatures (T_s). (b) MODIS T_s from 26 March, 2009 0326 UTC. Data gaps (white areas) result from the MODIS cloud mask. Black line denotes the position of the KT15 transect shown in (a). Gray triangles mark the position of the aerial photographs (P) 1-3. Legs (L) 1-3 denote the profiles shown in (d). (c) Aerial photographs taken parallel to the KT15 data. The picture size is 120 m × 80 m. (d) KT15 T_s across-polynya profiles. © Envisat ASAR image: T. Krumpen, AWI (2009).

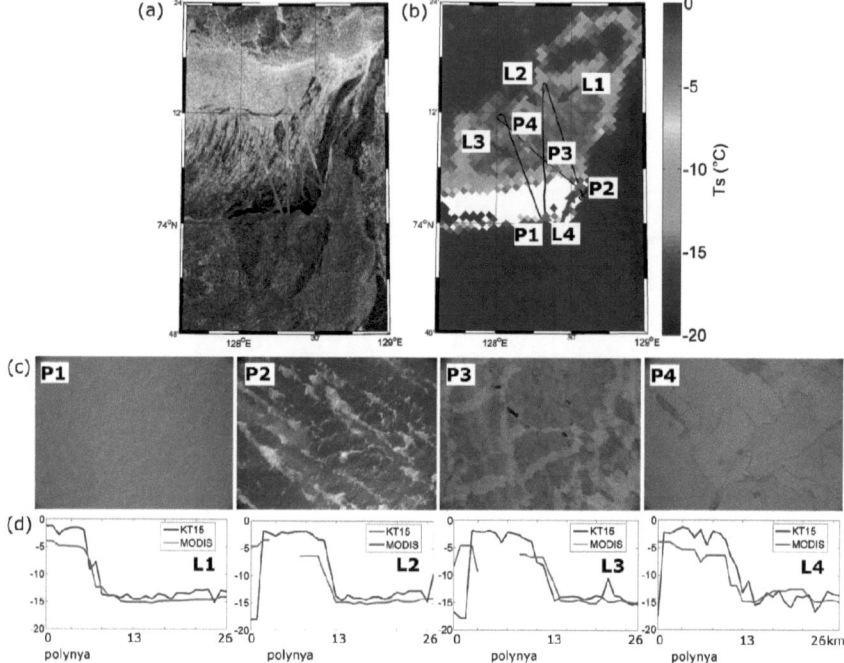

Figure A2: (a) Envisat ASAR image from 15 April, 2009 1232 UTC for illustration of the polynya conditions. Color-graded transect was measured at 14 April, 2009 around 0600 UTC and shows KT15 ice-surface temperatures (T_s). (b) MODIS T_s from 14 April, 2009 0445 UTC. Data gaps (white areas) result from the MODIS cloud mask. Black line denotes the position of the KT15 transect shown in (a). Gray triangles mark the position of the aerial photographs (P) 1-3. Legs (L) 1-3 denote the profiles shown in (d). (c) Aerial photographs taken parallel to the KT15 data. The picture size is 120 m × 80 m. (d) KT15 T_s across-polynya profiles. © Envisat ASAR image: T. Krumpen, AWI (2009).

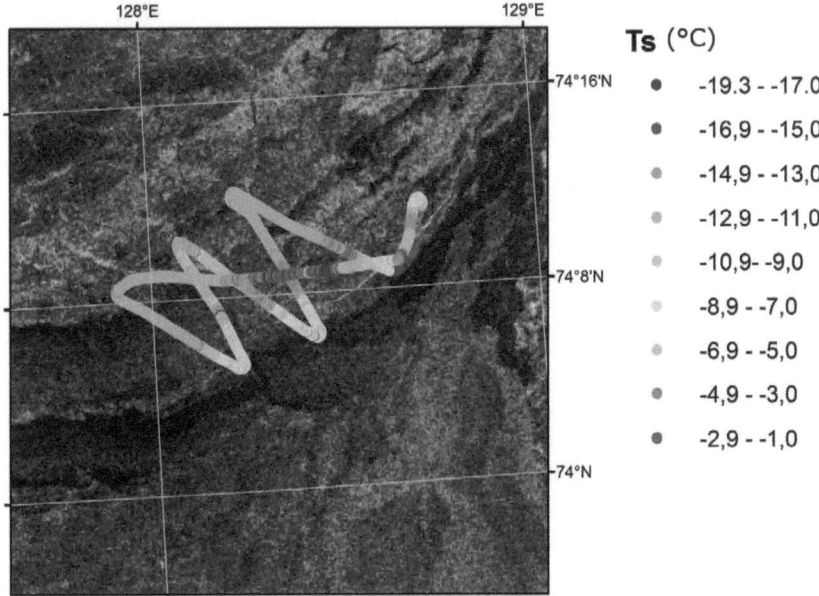

Figure A3: Envisat ASAR image from 27 March, 2009 1229 UTC for illustration of the polynya conditions. Color-graded transect was measured at 27 March, 2009 around 0630 UTC and shows KT15 ice-surface temperatures (T_s). © Envisat ASAR image: T. Krumpen, AWI (2009).

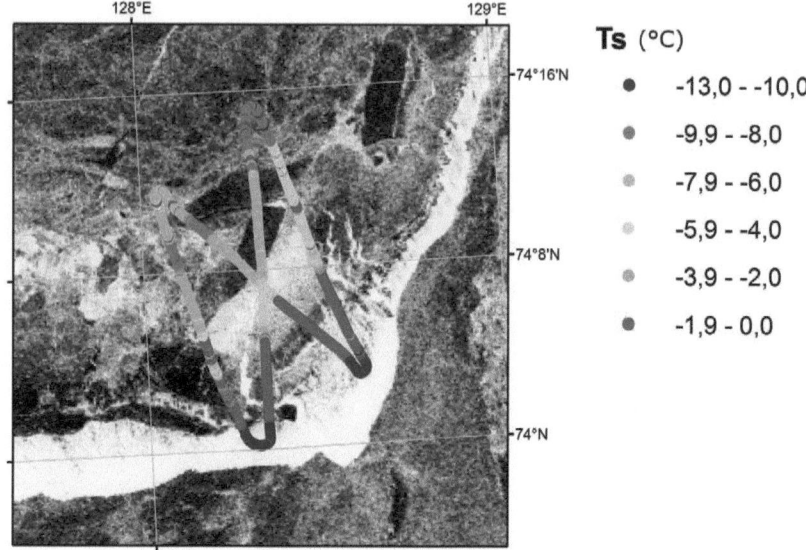

Figure A4: Envisat ASAR image from 8 April, 2009 0257 UTC for illustration of the polynya conditions. Color-graded transect was measured at 8 April, 2009 around 0550 UTC and shows KT15 ice-surface temperatures (T_s). © Envisat ASAR image: T. Krumpen, AWI (2009).

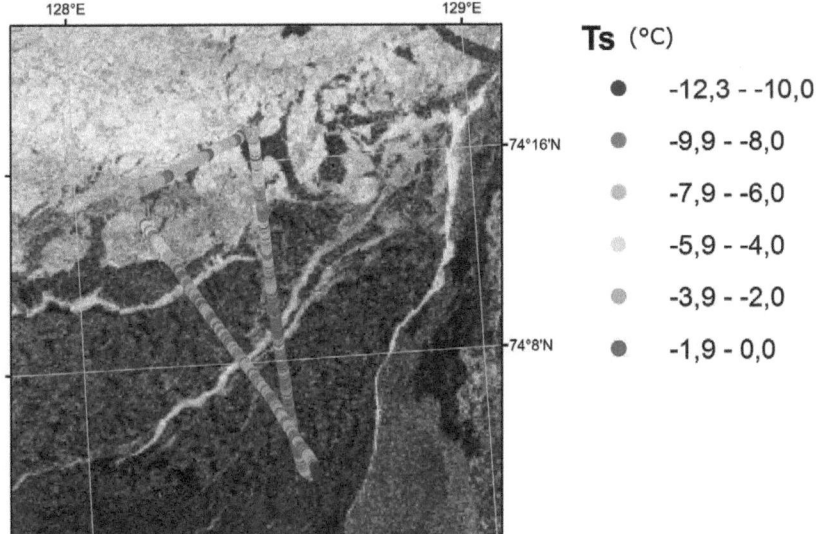

Figure A5: Envisat ASAR image from 21 April, 2009 0248 UTC for illustration of the polynya conditions. Color-graded transect was measured at 21 April, 2009 around 0500 UTC and shows KT15 ice-surface temperatures (T_s). © Envisat ASAR image: T. Krumpen, AWI (2009).

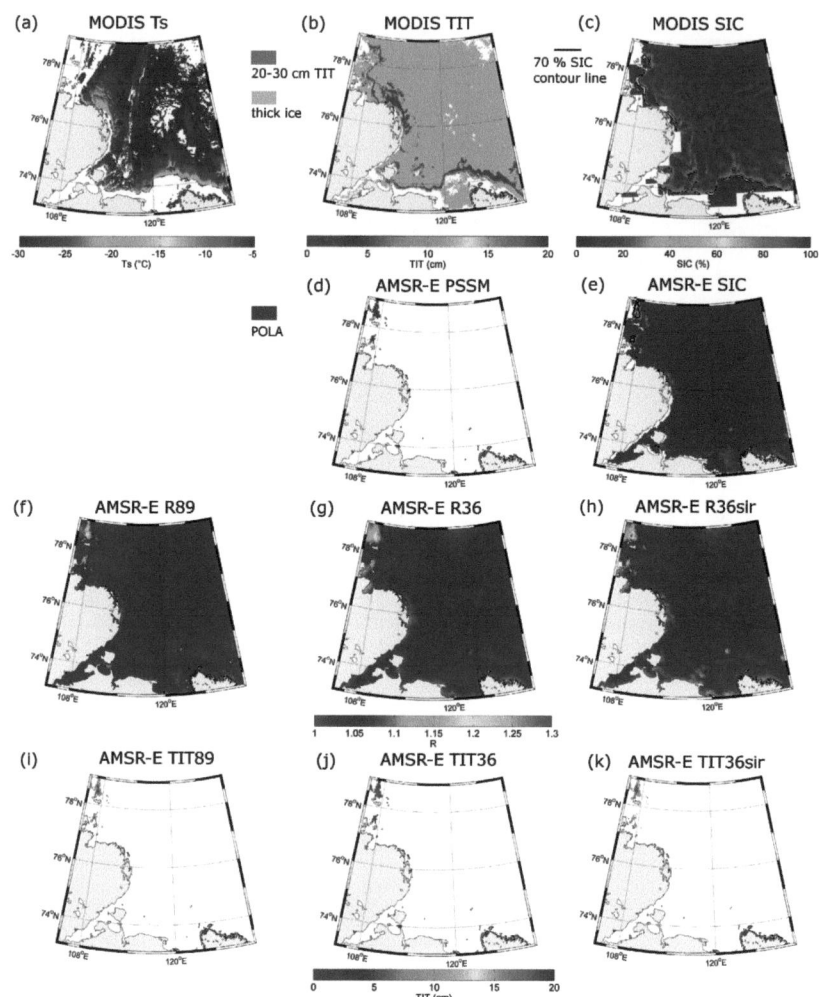

Figure A6: Overview of a various number of satellite data sets for the 26 December, 2007. (a) MODIS ice-surface temperature (T_s) measured at 0205 UTC. White areas are data gaps due to clouds. (b) Daily MODIS thin-ice thickness (TIT) map. (c) Daily MODIS sea-ice concentration (SIC). Black line denotes the 70 % SIC contour line. (d) Daily AMSR-E PSSM. (e) Daily AMSR-E SIC. Black line denotes the 70 % SIC contour line. (f) Daily AMSR-E polarization ratio of 89 GHz channels (R_{89}). (g) Daily AMSR-E R_{36}. (h) Daily AMSR-E R_{36sir}. (i) Daily AMSR-E thin-ice thickness calculated using 89 GHz channels (TIT_{89}). (j) Daily AMSR-E TIT_{36}. (k) Daily AMSR-E TIT_{36sir}. The TIT distributions in (i) – (k) are masked using PSSM polynya area. The MODIS data sets and Envisat ASAR data are reprojected to a polar stereographic 1 km × 1 km grid. All AMSR-E data sets are reprojected to a 6.25 km × 6.25 km grid.

Table A1: *The polynya area at the 26 December, 2007 is calculated using remote sensing data sets. Polynya area is derived from the thin-ice thickness using the 20 cm contour line as the polynya border. The 70 % threshold is applied for the calculation of polynya area from the sea-ice concentration.*

Data set	Polynya area (km^2)
Daily MODIS thin-ice thickness composite	13,209
Daily MODIS sea-ice concentration	4894
Daily AMSR-E sea-ice concentration	130
Daily PSSM polynya area	778

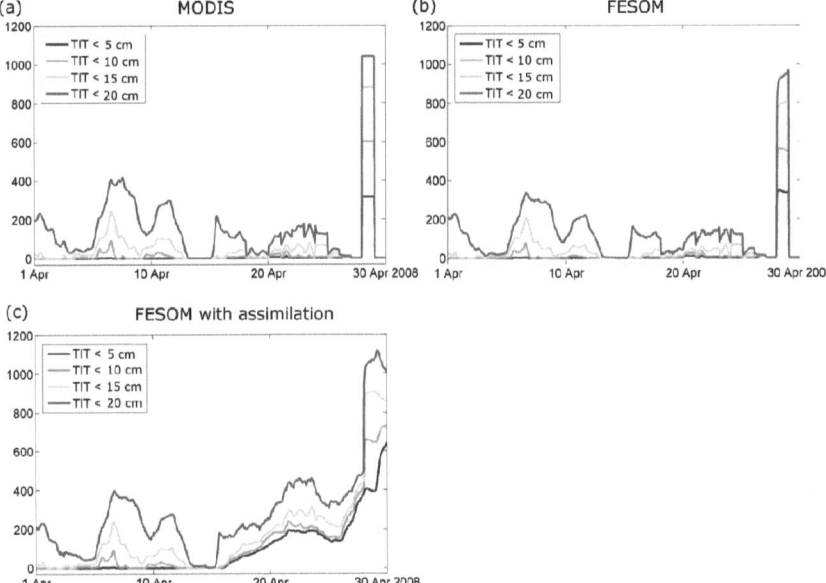

Figure A7: *First results of the combined remote-sensing assimilation method for April, 2008. Number of grid points of different ice-thickness classes for MODIS data, the independent FESOM simulations and the FESOM simulations with assimilated MODIS ice thickness data for April, 2008. (a) MODIS thin-ice thickness data. The data is interpolated to FESOM's 5-km grid. (b) Results of FESOM's independent model run. Only grid cells with available MODIS data are used. (c) Results of FESOM's assimilated model run. All grid cells are shown. Data sets are provided by D. Schröder, personal communication (2012).*

Acknowledgements

Working as a Ph.D. student for the Department of Environmental Meteorology, University of Trier was a challenging experience for me. In the last four years, many people have directly and indirectly influenced my research work. I owe many thanks to all these people for their help and encouragement.

First of all, I would like to express my gratitude to my supervisor Univ.-Prof. Dr. Günther Heinemann for his help and support. He always provided good ideas and inspiration. I would also like to thank Prof. Dr. Thomas Udelhoven for his willingness to serve as second reviewer for this thesis.

It is a pleasure to thank Dr. Sascha Willmes and Dr. David Schröder for providing encouraging and constructive feedback. Their insightful discussion and sustained support have contributed greatly to the success of this thesis.

I am deeply indebted to all my colleagues at the Department of Environmental Meteorology for their help and the great atmosphere.

I would also like to acknowledge Dr. Heidi Kassens, who enabled me to attend at the research expedition Transdrift XV in the Siberian Arctic. In addition, I would like to thank all the participants of the expedition, in particular, Prof. Dr. Alfred Helbig, Dr. Thomas Krumpen, Dr. Jens Hölemann and Torben Klagge for their support and help with collecting data.

I would like to thank Dr. Ralph Timmermann, Prof. Dr. Rüdiger Gerdes and Polona Itkin for providing model data. They always supported me when I had questions or needed advice.

I would like to acknowledge Prof. Dr. Christian Haas who gave me the change to observe scientific work at the Earth and Atmospheric Sciences Department, University of Alberta. I learned a lot there and gained new ideas for my research. The friendly welcome, the kind help and the enjoyable conversations made these two months a pleasure. I am also very grateful to Justin Beckers and Alec Casey for their support during the research stay and for proofreading the thesis.

This Ph.D. thesis is partly funded by the German Ministry for Education and Research (BMBF). The project 'System Laptev Sea' has the grant no. 03G0759D. Furthermore, I am very grateful to the Stipendienstiftung des Landes Rheinland-Pfalz for a two-year Ph.D. fellowship. A big thank to the German Academic Exchange Service (DAAD) and the Association of Polar Early Career Scientists (APECS) for travel funding, which has enabled me to participate in international conferences.

I am grateful to the institutes and organizations that granted me free access to data sets. Thanks to NASA (Washington, D.C., USA), NSIDC (Boulder, USA), NCEP (Boulder, USA) and ESA (Paris, France).

Last but not least, I owe a very big thank to my parents, Cäcilia and Egon, my siblings, Sabine and Michael and my friends, in particular, Anica, Christiane and Kerstin, for all their support and patience.

I want morebooks!

Buy your books fast and straightforward online - at one of the world's fastest growing online book stores! Environmentally sound due to Print-on-Demand technologies.

Buy your books online at

www.get-morebooks.com

Kaufen Sie Ihre Bücher schnell und unkompliziert online – auf einer der am schnellsten wachsenden Buchhandelsplattformen weltweit! Dank Print-On-Demand umwelt- und ressourcenschonend produziert.

Bücher schneller online kaufen

www.morebooks.de

VDM Verlagsservicegesellschaft mbH
Heinrich-Böcking-Str. 6-8
D - 66121 Saarbrücken   Telefax: +49 681 93 81 567-9

info@vdm-vsg.de
www.vdm-vsg.de

Printed by Books on Demand GmbH, Norderstedt / Germany